Integrated Optomechanical Analysis

Tutorial Texts Series

- *Integrated Optomechanical Analysis,* Keith B. Doyle, Victor L. Genberg, Gregory J. Michels, Vol. TT58
- *Thin-Film Design: Modulated Thickness and Other Stopband Design Methods,* Bruce Perilloux, Vol. TT57
- *Optische Grundlagen für Infrarotsysteme,* Max J. Riedl, Vol. TT56
- *An Engineering Introduction to Biotechnology,* J. Patrick Fitch, Vol. TT55
- *Image Performance in CRT Displays,* Kenneth Compton, Vol. TT54
- *Introduction to Laser Diode-Pumped Solid State Lasers,* Richard Scheps, Vol. TT53
- *Modulation Transfer Function in Optical and Electro-Optical Systems,* Glenn D. Boreman, Vol. TT52
- *Uncooled Thermal Imaging Arrays, Systems, and Applications,* Paul W. Kruse, Vol. TT51
- *Fundamentals of Antennas,* Christos G. Christodoulou and Parveen Wahid, Vol. TT50
- *Basics of Spectroscopy,* David W. Ball, Vol. TT49
- *Optical Design Fundamentals for Infrared Systems, Second Edition,* Max J. Riedl, Vol. TT48
- *Resolution Enhancement Techniques in Optical Lithography,* Alfred Kwok-Kit Wong, Vol. TT47
- *Copper Interconnect Technology,* Christoph Steinbrüchel and Barry L. Chin, Vol. TT46
- *Optical Design for Visual Systems,* Bruce H. Walker, Vol. TT45
- *Fundamentals of Contamination Control,* Alan C. Tribble, Vol. TT44
- *Evolutionary Computation: Principles and Practice for Signal Processing,* David Fogel, Vol. TT43
- *Infrared Optics and Zoom Lenses,* Allen Mann, Vol. TT42
- *Introduction to Adaptive Optics,* Robert K. Tyson, Vol. TT41
- *Fractal and Wavelet Image Compression Techniques,* Stephen Welstead, Vol. TT40
- *Analysis of Sampled Imaging Systems,* R. H. Vollmerhausen and R. G. Driggers, Vol. TT39
- *Tissue Optics: Light Scattering Methods and Instruments for Medical Diagnosis,* Valery Tuchin, Vol. TT38
- *Fundamentos de Electro-Óptica para Ingenieros,* Glenn D. Boreman, translated by Javier Alda, Vol. TT37
- *Infrared Design Examples,* William L. Wolfe, Vol. TT36
- *Sensor and Data Fusion Concepts and Applications, Second Edition,* L. A. Klein, Vol. TT35
- *Practical Applications of Infrared Thermal Sensing and Imaging Equipment, Second Edition,* Herbert Kaplan, Vol. TT34
- *Fundamentals of Machine Vision,* Harley R. Myler, Vol. TT33
- *Design and Mounting of Prisms and Small Mirrors in Optical Instruments,* Paul R. Yoder, Jr., Vol. TT32
- *Basic Electro-Optics for Electrical Engineers,* Glenn D. Boreman, Vol. TT31
- *Optical Engineering Fundamentals,* Bruce H. Walker, Vol. TT30
- *Introduction to Radiometry,* William L. Wolfe, Vol. TT29
- *Lithography Process Control,* Harry J. Levinson, Vol. TT28
- *An Introduction to Interpretation of Graphic Images,* Sergey Ablameyko, Vol. TT27
- *Thermal Infrared Characterization of Ground Targets and Backgrounds,* P. Jacobs, Vol. TT26
- *Introduction to Imaging Spectrometers,* William L. Wolfe, Vol. TT25
- *Introduction to Infrared System Design,* William L. Wolfe, Vol. TT24
- *Introduction to Computer-based Imaging Systems,* D. Sinha, E. R. Dougherty, Vol. TT23
- *Optical Communication Receiver Design,* Stephen B. Alexander, Vol. TT22
- *Mounting Lenses in Optical Instruments,* Paul R. Yoder, Jr., Vol. TT21
- *Optical Design Fundamentals for Infrared Systems,* Max J. Riedl, Vol. TT20
- *An Introduction to Real-Time Imaging,* Edward R. Dougherty, Phillip A. Laplante, Vol. TT19
- *Introduction to Wavefront Sensors,* Joseph M. Geary, Vol. TT18
- *Integration of Lasers and Fiber Optics into Robotic Systems,* J. A. Marszalec, E. A. Marszalec, Vol. TT17
- *An Introduction to Nonlinear Image Processing,* E. R. Dougherty, J. Astola, Vol. TT16
- *Introduction to Optical Testing,* Joseph M. Geary, Vol. TT15
- *Image Formation in Low-Voltage Scanning Electron Microscopy,* L. Reimer, Vol. TT12
- *Diazonaphthoquinone-based Resists,* Ralph Dammel, Vol. TT11

Integrated Optomechanical Analysis

Keith B. Doyle
Victor L. Genberg
Gregory J. Michels

Tutorial Texts in Optical Engineering
Volume TT58

Arthur R. Weeks, Jr., Series Editor
Invivo Research Inc. and University of Central Florida

SPIE PRESS
A Publication of SPIE—The International Society for Optical Engineering
Bellingham, Washington USA

Library of Congress Cataloging-in-Publication Data

Doyle, Keith B.
Integrated optomechanical analysis / Keith B. Doyle, Victor L. Genberg, Gregory J. Michels
p. cm. — (SPIE tutorial texts ; v. TT 58)
Includes bibliographical references and index.
ISBN 0-8194-4609-2 (softcover)
1. Optical instruments–Design and construction. I. Genberg, Victor L. II. Michels, Gregory J. III. Title. IV. Tutorial texts in optical engineering; v. TT58

TS513 .D69 2002
681'.4—dc21 2002067188
CIP

Published by

SPIE—The International Society for Optical Engineering
P.O. Box 10
Bellingham, Washington 98227-0010 USA
Phone: 360.676.3290
Fax: 360.647.1445
Email: spie@spie.org
www.spie.org

Printed in the United States of America.

Introduction to the Series

Since its conception in 1989, the Tutorial Texts series has grown to more than 60 titles covering many diverse fields of science and engineering. When the series was started, the goal of the series was to provide a way to make the material presented in SPIE short courses available to those who could not attend, and to provide a reference text for those who could. Many of the texts in this series are generated from notes that were presented during these short courses. But as stand-alone documents, short course notes do not generally serve the student or reader well. Short course notes typically are developed on the assumption that supporting material will be presented verbally to complement the notes, which are generally written in summary form to highlight key technical topics and therefore are not intended as stand-alone documents. Additionally, the figures, tables, and other graphically formatted information accompanying the notes require the further explanation given during the instructor's lecture. Thus, by adding the appropriate detail presented during the lecture, the course material can be read and used independently in a tutorial fashion.

What separates the books in this series from other technical monographs and textbooks is the way in which the material is presented. To keep in line with the tutorial nature of the series, many of the topics presented in these texts are followed by detailed examples that further explain the concepts presented. Many pictures and illustrations are included with each text and, where appropriate, tabular reference data are also included.

The topics within the series have grown from the initial areas of geometrical optics, optical detectors, and image processing to include the emerging fields of nanotechnology, biomedical optics, and micromachining. When a proposal for a text is received, each proposal is evaluated to determine the relevance of the proposed topic. This initial reviewing process has been very helpful to authors in identifying, early in the writing process, the need for additional material or other changes in approach that would serve to strengthen the text. Once a manuscript is completed, it is peer reviewed to ensure that chapters communicate accurately the essential ingredients of the processes and technologies under discussion.

It is my goal to maintain the style and quality of books in the series, and to further expand the topic areas to include new emerging fields as they become of interest to our reading audience.

Arthur R. Weeks, Jr.
Invivo Research Inc. and University of Central Florida

CONTENTS

Introduction

The numerical simulation of optical performance is typically a multidisciplinary effort comprising thermal, structural, and optical analysis tools. Performing detailed design trades such as computing optical performance as a function of temperature and mounting configuration requires the passing of data between the various models. In addition, it is common to develop finite element models to determine the thermal and structural response of an optical instrument due to environmental factors. The goals of this text are twofold; the first is to present finite element modeling techniques specific to optical systems and second, to present methods to integrate the thermal and structural response quantities into the optical model for detailed performance predictions.

The first two chapters provide a review and act as reference material for the rest of the chapters. The first chapter reviews mechanical engineering basics and finite element theory. Included in this section are the equations of elasticity, fracture mechanics, failure theories, heat transfer, structural dynamics, and a discussion on finite element modeling issues. The second chapter discusses optical fundamentals, optical performance metrics, and image formation. Included are discussions on polarized light, wavefront error, diffraction, the point spread function, and the modulation transfer function. Also presented is the use of orthogonal polynomials such as the Zernike polynomials to represent optical surface data.

Finite element model construction and analysis methods for predicting displacements of optical elements and support structures are discussed in Chapter 3. Topics include modeling methods for individual optical components, adhesive bond models, surface coating effects, flexure mounts, test supports, and assembly processes. The next two chapters discuss methods of integrating structural and thermal response quantities into the optical model and their effects on optical performance. Chapter 4 presents methods to integrate rigid-body errors and optical surface deformations, predict optical errors due to stress birefringence, compute line-of-sight jitter, and predict the effect mechanical obscurations have on image quality. Optothermal analysis methods are discussed in Chapter 5, including thermo-elastic and thermo-optic modeling techniques. Also discussed are methods to model bulk volumetric absorption, map temperatures from the thermal to structural model, and to account for moisture effects and adhesive curing.

Chapter 6 provides an introduction to the analysis of adaptive optics. Concepts and definitions including correctability and influence functions are discussed. Also, the mathematics to compute actuator inputs to minimize optical surface deformations are presented along with examples. Chapter 7 discusses structural optimization theory and applications, including the use of optical responses as constraints in structural optimization models, and an application of multidisciplinary optimization is reviewed. In Chapter 8, a simple telescope serves as an example for many of the analysis techniques discussed in earlier chapters. Model and results files are included in electronic format on a CD so that readers can review specific details of input and output and even run the example cases as desired. In Chapter 9, an integrated optomechanical analysis of a lens assembly is presented. Thermal, structural, and optical analyses are demonstrated to compute the change in focus, wavefront error, the point spread function, and the modulation transfer function as a function of laser power.

Keith B. Doyle
Victor L. Genberg
Gregory J. Michels
September 2002

Integrated Optomechanical Analysis

≺Chapter 1≻ Introduction to Mechanical Analysis Using Finite Elements

1.1 Integrated Optomechanical Analysis Issues

1.1.1 Integration issues

The optical performance of telescopes, lens barrels, and other optical systems are heavily influenced by mechanical effects. Figure 1.1 depicts the interaction between thermal, structural, and optical analysis. Each analysis type has its own specialized software to solve its own field specific problems. To predict interdisciplinary behavior, data must be passed between analysis types. In this book, emphasis is placed on the interaction of the three analysis disciplines.

1.1.2 Example: orbiting telescope

A simple finite element structural model of an orbiting telescope is shown in Fig. 1.2 and a corresponding optical model is shown in Fig. 1.3. Because of dynamic disturbances, the optics may move relative to each other and distort elastically. From the finite element model, the motions of each node point are predicted. To determine the effect on optical performance, it is necessary to pass the data to the optical analysis program in a form that is acceptable. This usually requires a special post-processing program as described in later chapters. Typically, the structural data must be converted to the optical coordinate system, optical units, and sign convention, then fit with Zernike polynomials or interpolated to interferogram arrays (Chapter 4).

To create a valid and accurate structural model, the analyst must be aware of modeling techniques for mirrors, mounts, and adhesive bonds (Chapter 3). Incorporating image-motion equations inside the FE model (Chapter 4) allows for image-motion output directly from a vibration analysis. The vibrations may be due to harmonic or random loads. To determine if a mirror will fracture, the analyst must understand the type of failure analysis required and how to search over load envelopes. During the processing (grinding, polishing, and coating) of a mirror, it may be tested under various support conditions that require their own analysis (Chapter 3). Analysis of the assembly process (Chapter 3) will predict locked-in strains and create an optical back out that can be factored into the overall system performance. Performance of the flexible primary mirror can be improved by adding actuators and sensors to create an adaptive mirror (Chapter 6). Using optimum design techniques (Chapter 7), the design can be made more efficient and robust. The specific details of the analyses on this telescope are demonstrated in Chapter 8.

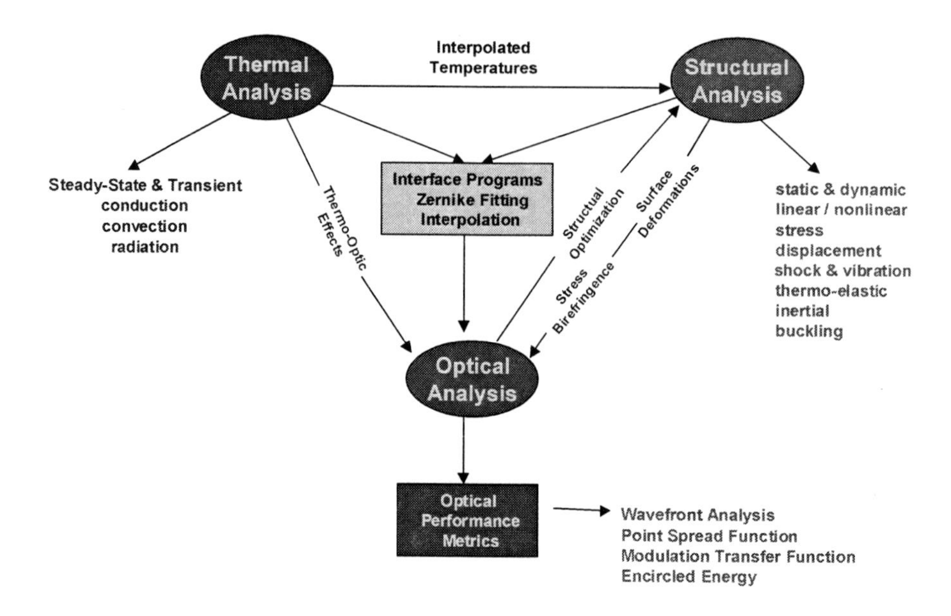

Figure 1.1 Optomechanical analysis interaction.

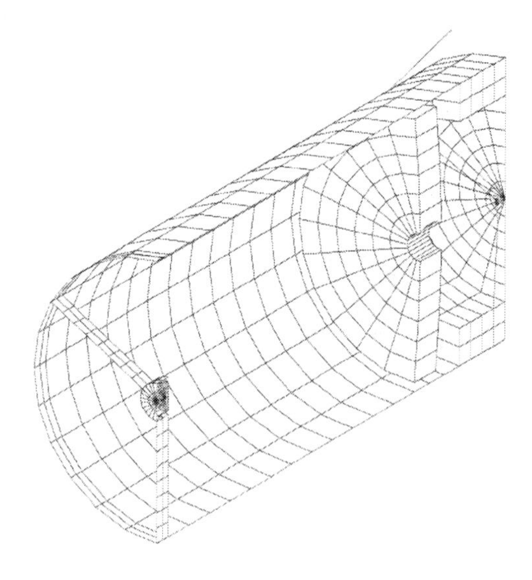

Figure 1.2 Telescope structural analysis model.

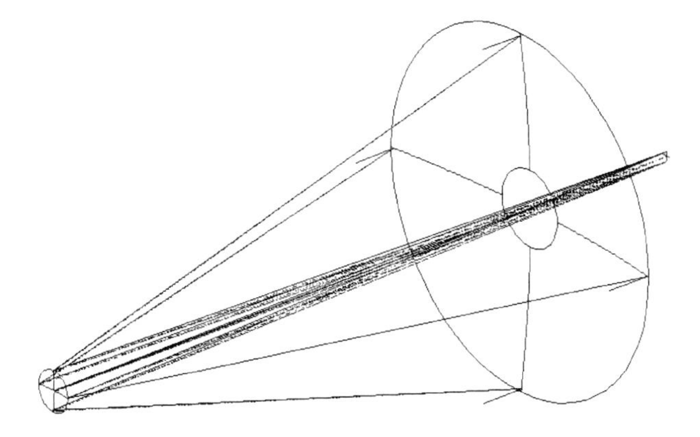

Figure 1.3 Telescope optical analysis model.

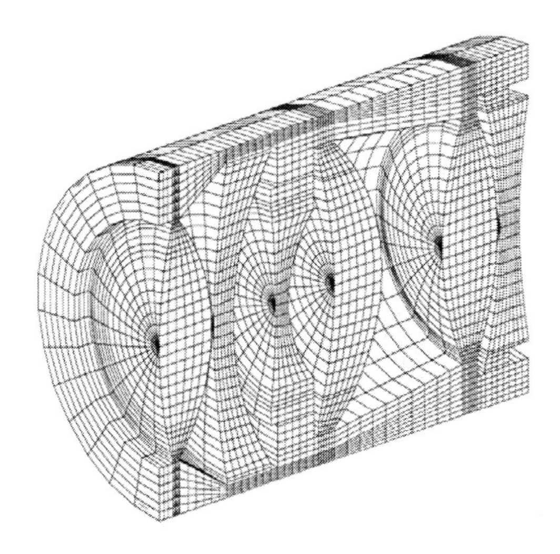

Figure 1.4 Lens barrel structural model.

1.1.3 Example: lens barrel

The lens barrel in Fig. 1.4 is representative of components used in a variety of applications from optical lithography to projection systems. Often the optical beam causes thermal loading on the lenses. Analyzing for the steady-state or transient temperature distribution is the first analysis required (Chapter 5). The resulting temperature profiles may cause an optical index change throughout each lens, which affects the optical performance (Chapter 5). As part of the structural analysis, temperatures must be applied that will require interpolation (Chapter 5) if the structural model is different from the thermal model. The thermoelastic stresses cause distortion (Chapter 3), and may cause stress birefringence effects (Chapter 4). Each of these effects requires special software to analyze the FEA results and present the data in a format suitable for optical programs. If the structure and loading have symmetry, techniques can be used to reduce the computation required. The example lens barrel in Chapter 9 demonstrates many of the techniques discussed throughout the text.

1.2 Elasticity Review

1.2.1 Three-dimensional elasticity

TERMINOLOGY:
E = Young's modulus = slope of stress-strain curve
ν = Poisson's ratio = contraction in y, z due to elongation in x
α = Coefficient of thermal expansion (CTE)
σ = Stress = force/unit area
u, v, w = Displacements in x, y, z directions
e = Total strain = $\delta u/\delta x$ = stretch/unit length = $\varepsilon + e_T$
ε = Mechanical strain = due to applied stress
e_T = Thermal strain = due to temperature change $\Delta T = \alpha \Delta T$

Stress components are shown in Fig. 1.5. The strain-component notation is analogous to the stress notation. Pictorially represented in Fig. 1.6, the strain-displacement relations are

$$\begin{aligned} \varepsilon_x &= du/dx & \gamma_{xy} &= [du/dy + dv/dx] \\ \varepsilon_y &= dv/dx & \gamma_{yz} &= [dv/dz + dw/dy]. \\ \varepsilon_z &= dw/dx & \gamma_{zx} &= [dw/dx + du/dz] \end{aligned} \tag{1.1}$$

Shear strain may be defined as above (as in MSC/Nastran[1]) or as half of that value (as in MSC/Patran[1]). The engineer must be aware of which definition is used in the analysis software.

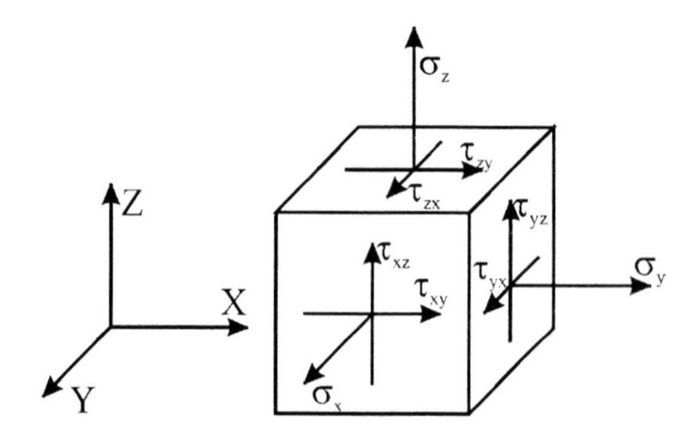

Figure 1.5 Stress components.

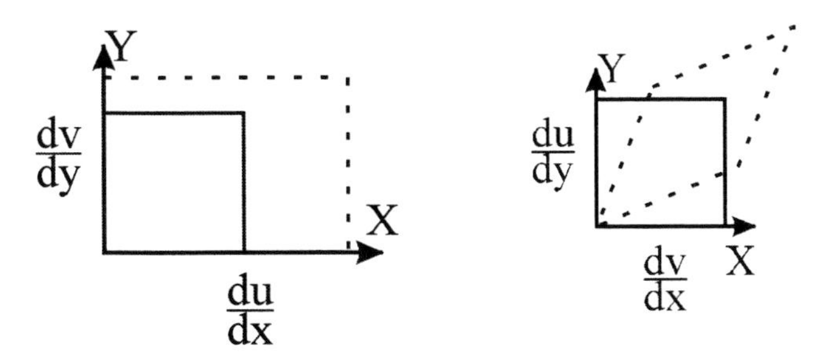

Figure 1.6 Strain-displacement relations.

For isotropic materials, the full 3D stress-strain relations may be represented as

$$\begin{Bmatrix} e_x \\ e_y \\ e_z \\ e_{xy} \\ e_{yz} \\ e_{zx} \end{Bmatrix} = \frac{1}{E}\begin{bmatrix} 1 & -\nu & -\nu & 0 & 0 & 0 \\ -\nu & 1 & -\nu & 0 & 0 & 0 \\ -\nu & -\nu & 1 & 0 & 0 & 0 \\ 0 & 0 & 0 & 2(1+\nu) & 0 & 0 \\ 0 & 0 & 0 & 0 & 2(1+\nu) & 0 \\ 0 & 0 & 0 & 0 & 0 & 2(1+\nu) \end{bmatrix}\begin{Bmatrix} \sigma_x \\ \sigma_y \\ \sigma_z \\ \tau_{xy} \\ \tau_{yz} \\ \tau_{zx} \end{Bmatrix} + \alpha\Delta T\begin{Bmatrix} 1 \\ 1 \\ 1 \\ 0 \\ 0 \\ 0 \end{Bmatrix}, \tag{1.2}$$

or in inverted form:

$$\begin{Bmatrix} \sigma_x \\ \sigma_y \\ \sigma_z \\ \tau_{xy} \\ \tau_{yz} \\ \tau_{zx} \end{Bmatrix} = \frac{E}{(1+\nu)(1-2\nu)}\begin{bmatrix} 1-\nu & \nu & \nu & 0 & 0 & 0 \\ \nu & 1-\nu & \nu & 0 & 0 & 0 \\ \nu & \nu & 1-\nu & 0 & 0 & 0 \\ 0 & 0 & 0 & \frac{1-2\nu}{2} & 0 & 0 \\ 0 & 0 & 0 & 0 & \frac{1-2\nu}{2} & 0 \\ 0 & 0 & 0 & 0 & 0 & \frac{1-2\nu}{2} \end{bmatrix}\begin{Bmatrix} e_x \\ e_y \\ e_z \\ e_{xy} \\ e_{yz} \\ e_{zx} \end{Bmatrix} - \frac{E\alpha\Delta T}{1\text{-}2\nu}\begin{Bmatrix} 1 \\ 1 \\ 1 \\ 0 \\ 0 \\ 0 \end{Bmatrix}. \tag{1.3}$$

The form in Eq. (1.2) is more intuitive since one can see how applied stress causes strain effects. However, the form in Eq. (1.3) is commonly used in FEA programs. The coefficient matrix in Eq. (1.3) is often referred to as the material matrix. If the material is orthotropic, then the stress-strain relations are represented as

$$\begin{Bmatrix} e_x \\ e_y \\ e_z \\ e_{xy} \\ e_{yz} \\ e_{zx} \end{Bmatrix} = \begin{bmatrix} \frac{1}{E_x} & -\frac{\nu_{yx}}{E_y} & -\frac{\nu_{zx}}{E_z} & 0 & 0 & 0 \\ -\frac{\nu_{xy}}{E_x} & \frac{1}{E_y} & -\frac{\nu_{zy}}{E_z} & 0 & 0 & 0 \\ -\frac{\nu_{xz}}{E_x} & -\frac{\nu_{yz}}{E_y} & \frac{1}{E_z} & 0 & 0 & 0 \\ 0 & 0 & 0 & \frac{1}{G_{xy}} & 0 & 0 \\ 0 & 0 & 0 & 0 & \frac{1}{G_{yz}} & 0 \\ 0 & 0 & 0 & 0 & 0 & \frac{1}{G_{zx}} \end{bmatrix} \begin{Bmatrix} \sigma_x \\ \sigma_y \\ \sigma_z \\ \tau_{xy} \\ \tau_{yz} \\ \tau_{zx} \end{Bmatrix} + \Delta T \begin{Bmatrix} \alpha_x \\ \alpha_y \\ \alpha_z \\ 0 \\ 0 \\ 0 \end{Bmatrix}. \quad (1.4)$$

$$\begin{Bmatrix} \sigma_x \\ \sigma_y \\ \sigma_z \\ \tau_{xy} \\ \tau_{yz} \\ \tau_{zx} \end{Bmatrix} = \begin{bmatrix} \frac{1-\nu_{yz}\nu_{zy}}{\Psi}E_x & \frac{\nu_{xy}+\nu_{zy}\nu_{xz}}{\Psi}E_y & \frac{\nu_{xz}+\nu_{xy}\nu_{yz}}{\Psi}E_z & 0 & 0 & 0 \\ \frac{\nu_{yx}+\nu_{zx}\nu_{yz}}{\Psi}E_x & \frac{1-\nu_{xz}\nu_{zx}}{\Psi}E_y & \frac{\nu_{yz}+\nu_{yx}\nu_{xz}}{\Psi}E_z & 0 & 0 & 0 \\ \frac{\nu_{zx}+\nu_{yx}\nu_{zy}}{\Psi}E_x & \frac{\nu_{zy}+\nu_{xy}\nu_{zx}}{\Psi}E_y & \frac{1-\nu_{xy}\nu_{yx}}{\Psi}E_z & 0 & 0 & 0 \\ 0 & 0 & 0 & G_{xy} & 0 & 0 \\ 0 & 0 & 0 & 0 & G_{yz} & 0 \\ 0 & 0 & 0 & 0 & 0 & G_{zx} \end{bmatrix} \begin{Bmatrix} e_x \\ e_y \\ e_z \\ e_{xy} \\ e_{yz} \\ e_{zx} \end{Bmatrix} - \Delta T \begin{Bmatrix} \alpha_x \\ \alpha_y \\ \alpha_z \\ 0 \\ 0 \\ 0 \end{Bmatrix},$$

$$\Psi = 1 - \nu_{xy}\nu_{yx} - \nu_{yz}\nu_{zy} - \nu_{zx}\nu_{xz} - 2\nu_{yx}\nu_{zy}\nu_{xz}, \quad (1.5)$$

ν_{ij} is $-\varepsilon_j/\varepsilon_i$ for uniaxial stress σ_I.

The above equations may be used to analyze material that is orthotropic in nature, or they may be used to analyze isotropic materials that are fabricated by a method so they act in an orthotropic fashion (See Chapter 3).

1.2.2 Two-dimensional plane stress

Although all structures are truly 3D, it is computationally efficient to approximate thin structures (plates and shells) with 2D plane stress relations for isotropic materials [Eqs. (1.6) and (1.7)]. If a thin structure lies in the *X-Y* plane, then the normal (*Z*) stress components are assumed to be zero:

$$\sigma_z = \tau_{yz} = \tau_{zx} = 0$$

$$\begin{Bmatrix} e_x \\ e_y \\ e_{xy} \end{Bmatrix} = \frac{1}{E}\begin{bmatrix} 1 & -\nu & 0 \\ -\nu & 1 & 0 \\ 0 & 0 & 2(1+\nu) \end{bmatrix}\begin{Bmatrix} \sigma_x \\ \sigma_y \\ \tau_{xy} \end{Bmatrix} + \alpha\Delta T\begin{Bmatrix} 1 \\ 1 \\ 0 \end{Bmatrix}, \tag{1.6}$$

and

$$\begin{Bmatrix} \sigma_x \\ \sigma_y \\ \tau_{xy} \end{Bmatrix} = \frac{E}{1-\nu^2}\begin{bmatrix} 1 & \nu & 0 \\ \nu & 1 & 0 \\ 0 & 0 & \frac{1-\nu}{2} \end{bmatrix}\begin{Bmatrix} e_x \\ e_y \\ e_{xy} \end{Bmatrix} - \frac{E\alpha\Delta T}{1-\nu}\begin{Bmatrix} 1 \\ 1 \\ 0 \end{Bmatrix}. \tag{1.7}$$

Under this assumption, the normal strains are not zero but given as

$$\begin{aligned} e_z &= \frac{-\nu}{E}\left(\sigma_x + \sigma_y\right) + \alpha\ \Delta T \\ e_{yz} &= e_{zx} = 0 \end{aligned}, \tag{1.8}$$

Thus, in-plane stretching causes the material to get thinner.

For orthotropic materials, such as a graphite-epoxy panel, the plane stress relations are given as

$$\begin{Bmatrix} e_x \\ e_y \\ e_{xy} \end{Bmatrix} = \begin{bmatrix} \frac{1}{E_x} & -\frac{\nu_{yx}}{E_y} & 0 \\ -\frac{\nu_{xy}}{E_x} & \frac{1}{E_y} & 0 \\ 0 & 0 & \frac{1}{G_{xy}} \end{bmatrix}\begin{Bmatrix} \sigma_x \\ \sigma_y \\ \tau_{xy} \end{Bmatrix} + \Delta T\begin{Bmatrix} \alpha_x \\ \alpha_y \\ 0 \end{Bmatrix}, \tag{1.9}$$

and

$$\begin{Bmatrix} \sigma_x \\ \sigma_y \\ \tau_{xy} \end{Bmatrix} = \begin{bmatrix} \frac{1}{1-\nu_{xy}\nu_{yx}}E_x & \frac{\nu_{xy}}{1-\nu_{xy}\nu_{yx}}E_y & 0 \\ \frac{\nu_{yx}}{1-\nu_{xy}\nu_{yx}}E_x & \frac{1}{1-\nu_{xy}\nu_{yx}}E_y & 0 \\ 0 & 0 & G_{xy} \end{bmatrix}\begin{Bmatrix} e_x & \alpha_x \\ e_y & -\Delta T\,\alpha_y \\ e_{xy} & 0 \end{Bmatrix}. \tag{1.10}$$

Through the thickness, strains are again nonzero:

$$e_z = -\frac{\nu_{xz}}{E_x}\sigma_x - \frac{\nu_{yz}}{E_y}\sigma_y + \alpha_z \ \Delta T \quad . \tag{1.11}$$
$$e_{yz} = e_{zx} = 0$$

1.2.3 Two-dimensional plane strain

An alternate approximation is to assume that the normal strains are 0, $\varepsilon_Z = \gamma_{yz} = \gamma_{zx} = 0$. This condition can occur for very wide, thin-bond areas, or for long (in *Z*) uniform structures. The isotropic plane-strain relations are

$$\begin{Bmatrix} e_x \\ e_y \\ e_{xy} \end{Bmatrix} = \frac{1+\nu}{E}\begin{bmatrix} 1-\nu & -\nu & 0 \\ -\nu & 1-\nu & 0 \\ 0 & 0 & 2 \end{bmatrix}\begin{Bmatrix} \sigma_x \\ \sigma_y \\ \tau_{xy} \end{Bmatrix} + (1+\nu)\alpha\Delta T\begin{Bmatrix} 1 \\ 1 \\ 0 \end{Bmatrix}, \tag{1.12}$$

and

$$\begin{Bmatrix} \sigma_x \\ \sigma_y \\ \tau_{xy} \end{Bmatrix} = \frac{E}{(1+\nu)(1-2\nu)}\begin{bmatrix} 1-\nu & \nu & 0 \\ \nu & 1-\nu & 0 \\ 0 & 0 & \frac{1-2\nu}{2} \end{bmatrix}\begin{Bmatrix} e_x \\ e_y \\ e_{xy} \end{Bmatrix} - \frac{E\alpha\Delta T}{1-2\nu}\begin{Bmatrix} 1 \\ 1 \\ 0 \end{Bmatrix}. \tag{1.13}$$

The normal stress is not zero in this assumption, but given as

$$\sigma_{zz} = \nu\left(\sigma_{xx} + \sigma_{yy}\right) - E\alpha\Delta T \ , \tag{1.14}$$
$$\tau_{yz} = \tau_{zx} = 0 \ .$$

To be complete, the orthotropic plane-strain relations are given as

$$\begin{Bmatrix} e_x \\ e_y \\ e_{xy} \end{Bmatrix} = \begin{bmatrix} \frac{1-\nu_{zx}\nu_{xz}}{E_x} & \frac{\nu_{yx}-\nu_{yz}\nu_{xz}}{E_y} & 0 \\ \frac{\nu_{xy}-\nu_{zx}\nu_{zy}}{E_x} & \frac{1-\nu_{yz}\nu_{zy}}{E_y} & 0 \\ 0 & 0 & \frac{1}{G_{xy}} \end{bmatrix}\begin{Bmatrix} \sigma_x \\ \sigma_y \\ \tau_{xy} \end{Bmatrix} + \Delta T\begin{Bmatrix} \alpha_x \\ \alpha_y \\ 0 \end{Bmatrix}, \tag{1.15}$$

and

$$\begin{Bmatrix} \sigma_x \\ \sigma_y \\ \tau_{xy} \end{Bmatrix} = \begin{bmatrix} \frac{1-\nu_{yz}\nu_{zy}}{\Psi} E_x & \frac{\nu_{xy}+\nu_{zy}\nu_{xz}}{\Psi} E_y & 0 \\ \frac{\nu_{yx}+\nu_{zx}\nu_{yz}}{\Psi} E_x & \frac{1-\nu_{xz}\nu_{zx}}{\Psi} E_y & 0 \\ 0 & 0 & G_{xy} \end{bmatrix} \begin{Bmatrix} e_x & \alpha_x \\ e_y & -\Delta T \alpha_y \\ e_{xy} & 0 \end{Bmatrix} . \quad (1.16)$$

1.2.4 Principal stress and equivalent stress

Stress failure cannot be determined directly from a general 2D or 3D state of stress. A general state of stress is processed to determine principal stresses or an equivalent stress, which is then used as a failure criterion. For a general 2D state of stress at a point (σ_x, σ_y, σ_{xy}), Mohr's circle (Fig. 1.7) is used to find the state of principal stress, which is defined as an orientation with *no* shear stress,

$(\sigma_1, \sigma_2, 0)$ where $C = \frac{\sigma_x + \sigma_y}{2}$ and $R = \sqrt{\tau_{xy}^{\ 2} + \left(\frac{\sigma_x - \sigma_y}{2}\right)^2}$

and $\sigma_1 = C+R$, $\sigma_2 = C-R$.

Ductile materials such as aluminum or steel follow the Maximum Distortion Energy Theory, in which yielding occurs when the Von Mises stress (σ_{vm}) from

$$\sigma_{vm} = \frac{1}{\sqrt{2}}\sqrt{(\sigma_1-\sigma_2)^2+(\sigma_2-\sigma_3)^2+(\sigma_3-\sigma_1)^2} \quad (1.17)$$

reaches the material yield stress.

Brittle materials such as common glasses follow fracture-mechanics laws in which fracture occurs when the stress-intensity factor (K) reaches the fracture toughness (K_C) of the material. K is computed from maximum principal stress, or maximum shear stress, flaw size, and geometry, and K_c is a material property. See Sec.1.5 for more details.

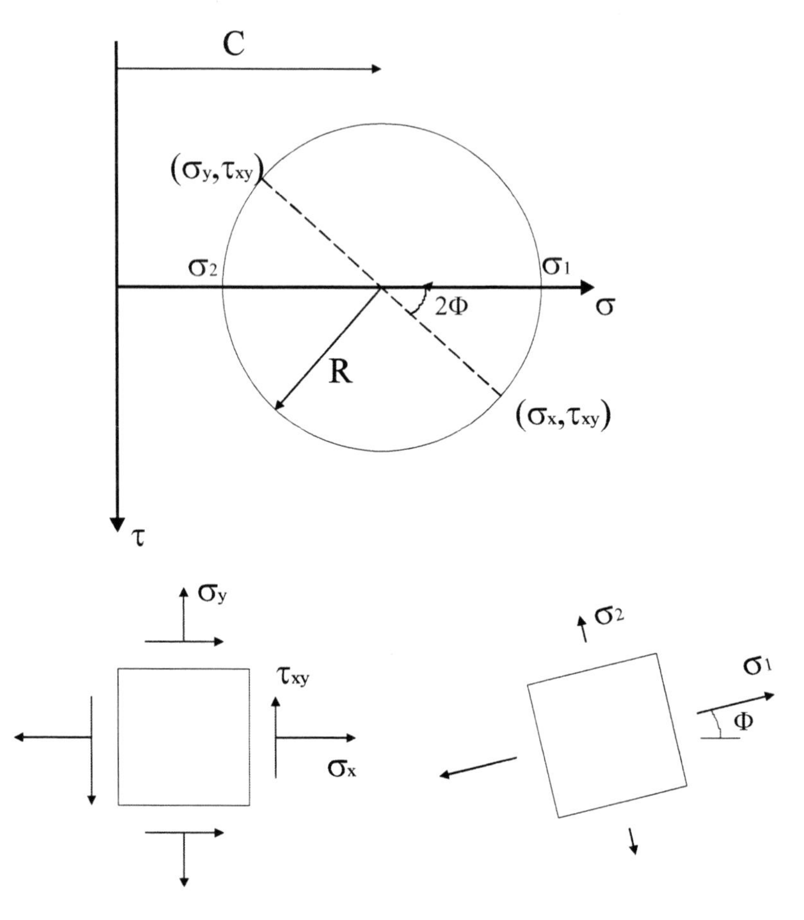

Figure 1.7 Mohr's Circle for 2D stress.

1.3 Basics of Finite Element Analysis

1.3.1 Finite element theory

FINITE DIFFERENCE (APPROXIMATE THE *MATH*):

- **⟨1⟩ Write equilibrium as the governing differential equation.**
- **⟨2⟩ Write derivatives as differences on a uniform grid.**
- **⟨3⟩ Solve the resulting matrix equation for behavior at the grid points.**
- **⟨4⟩ Odd-shaped boundaries are difficult to handle.**

FINITE ELEMENTS (APPROXIMATE THE *PHYSICS*):

- **⟨1⟩ Subdivide the body into simple elements of arbitrary size and shape.**
- **⟨2⟩ Assume simple polynomial behavior in each element.**
- **⟨3⟩ Write equilibrium at the nodes and solve for nodal values.**
- **⟨4⟩ Odd geometry is easily handled.**

Structure behavior in a continuous body is defined by differential equations, which are usually impossible to solve for real problems with complex geometry. Two common methods of approximation are finite difference and finite elements. This text concentrates on the finite element analysis (FEA) technique that is widely used in the analysis of optical structures.

In FEA, the displacement is assumed to have a simple polynomial behavior over an element. For the 1D truss element in Fig. 1.8, the assumed linear displacement is given by

$$u(x) = N_1\, U_1 + N_2\, U_2 = \Sigma\, N_j\, U_j = [N]\{U\},$$

$$N_1 = 1 - x/L\,,$$

and (1.18)

$$N_2 = x/L,$$

where U_j = displacement of node j (variable to be solved for), and N_j = shape function for node j (N_j = 1 at j, N_j = 0 at all other nodes). Thus, a continuous function, u(x), can be written in terms of discrete values, U_j. Using this relationship, stress and strain can also be written as a function of nodal variables U, as follows:

$$\varepsilon = du/dx = (d/dx)\, \Sigma\, N_j\, U_j = \Sigma\, dN_j/dx\, U_j = [B]\, \{U\},$$

$$\sigma = E\, \varepsilon = [G]\, [B]\, \{U\}. \qquad (1.19)$$

Potential energy Π is written as an integral over the element volume of the strain energy minus the work done, W_p, by the vector of applied nodal forces P:

$$\Pi = 0.5\int \varepsilon^T \sigma dV - W_P = 0.5\int U^T B^T G B U dV - U^T P\,. \qquad (1.20)$$

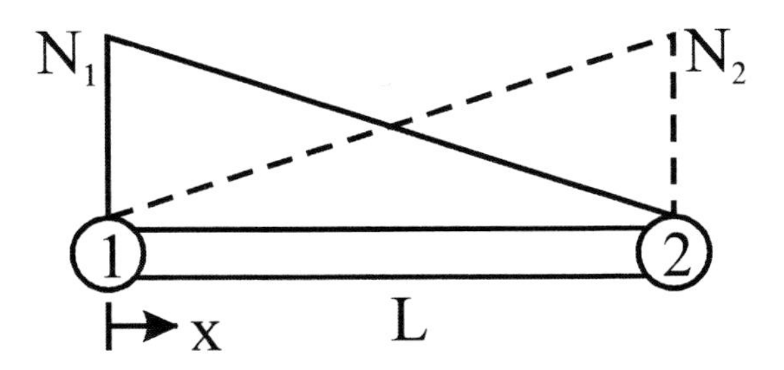

Figure 1.8 Linear shape functions.

Minimize Π with respect to the variables U = nodal displacements:

$$d\Pi / dU = 0 = \int B^T G B dV U - P = kU - P \ . \tag{1.21}$$

Thus, the element stiffness matrix [k] is:

$$k = \int B^T G B dV \ , \tag{1.22}$$

which, for the 1D truss element is

$$k = \frac{AE}{L} \begin{bmatrix} 1 & -1 \\ -1 & 1 \end{bmatrix} \ . \tag{1.23}$$

Generally, each element's stiffness matrix must be transformed into the global coordinate system used for nodal displacements via a coordinate transformation matrix, T:

$$k_g = T^T k T. \tag{1.24}$$

The element matrices are then assembled into system matrix K, resulting in the system level equilibrium equations:

$$[K] \{U\} = \{P\}. \tag{1.25}$$

After proper boundary conditions and loads are applied to the model, the above equations are solved for nodal displacements U. If desired, element stresses are determined from Eq. 1.19.

The same derivation may be applied to 2D plate and 3D solid elements if the shape functions add the appropriate spatial variables y and z. The order of the shape functions can be increased from linear with two nodes per edge, to quadratic with three nodes per edge, and higher. For additional information on finite element theory, see Refs. 2–4.

1.3.2 Element performance

The element-shape functions in the previous section determine the behavior and accuracy of the model. Consider the simple cantilever beam in Fig. 1.9, which is subject to a variety of load conditions. The load cases 1–3 exercise the membrane (in-plane) behavior, whereas cases 4–5 exercise bending (out-of-plane) behavior. The structure was modeled with a variety of 2D shell elements as shown in Fig. 1.10. From the results obtained from MSC/Nastran version 2001 listed in

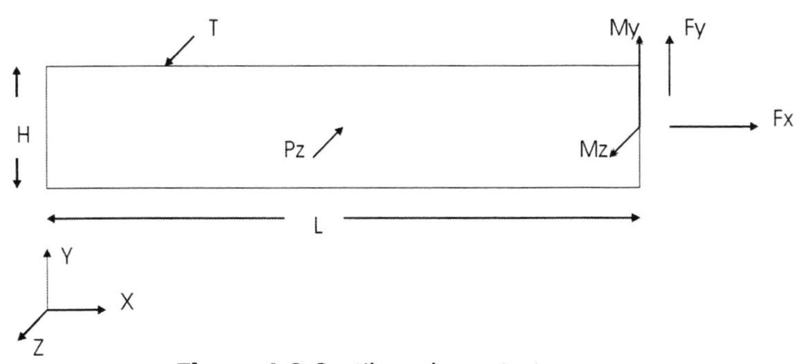

Figure 1.9 Cantilever beam test case.

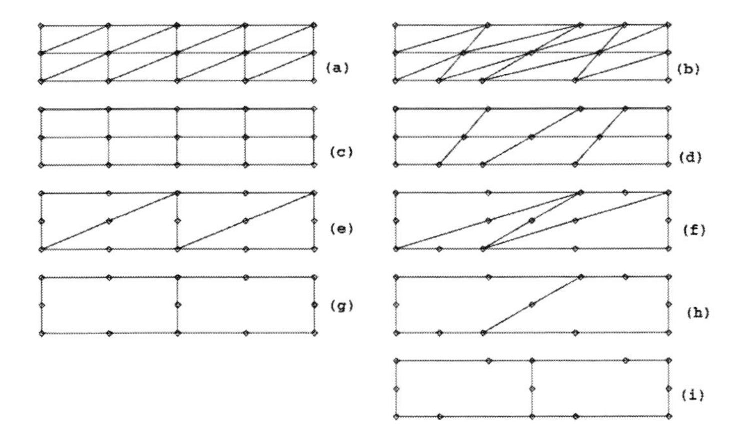

Figure 1.10 Two-dimensional plate-element models of cantilever beam.

Table 1.1 where the results are normalized by dividing by the exact value, the following conclusions can be drawn:

≺1≻ All elements correctly predict constant stress (case 1).
≺2≻ Tria3 is very poor for linear membrane stress (cases 2 and 3).
≺3≻ Other elements do well, even if distorted, for linear membrane stress.
≺4≻ All elements do well for plate bending.
≺5≻ Quadratic elements must have nodes located at the mid-point of the edges or accuracy degrades.

This is a simple test case, testing only a few of the capabilities of shell elements. Reference 4 lists many other test cases required to fully check element performance characteristics. There are many different shell-element formulations in the literature so other FEA programs may not obtain the same results as Table

1.1. The analyst is encouraged run similar test cases on the particular program of interest.

Table 1.1 Two-dimensional shell results for cantilever beam.

MEMBRANE (IN-PLANE) BEHAVIOR:

<1> Fx = axial load = uniform, constant stress

<2> Mz = moment in-plane = axial stress is linear in *y*

<3> Fy = shear force in-plane = axial stress linear in *x* and *y*

PLATE BENDING (OUT-OF-PLANE) BEHAVIOR:

<1> My = moment out-of-plane = stress constant in *x*

<2> pz = normal pressure = stress linear in *x*

MODELS:	Fx	Mz	Fy	My	pz
<a> Tria3–uniform	1.00	0.30	0.32	1.00	1.00
<b> Tria3–distorted	1.00	0.12	0.16	1.00	0.96
<c> Quad4–uniform	1.00	1.00	0.98	1.00	1.00
<d> Quad4–distorted	1.00	0.98[1]	0.96[2]	1.00	1.00
<e> Tria6–uniform	1.00	1.00	0.96	1.00	1.00
<f> Tria6–dististorted	1.00	1.00	0.82	1.00	0.84
<g> Quad8–uniform	1.00	1.00	1.00	1.00	1.00
<h> Quad8–distorted	1.00	0.98	0.94	0.91	0.83
<i> Quad8–midside-offset	1.00	0.59	0.59	0.43	0.44

1 = 0.41 prior to Version 70.7

2 = 0.47 prior to Version 70.7

FE is an approximate solution. As the FE mesh is made finer, more degrees-of-freedom (DOF) are added to the model and the approximation improves. For example, if the cantilever beam above is increased from 8 to 64 Tria3 elements, the tip displacement error is reduced from 70% to 10%. In FE theory, as the number of elements is increased, with their resulting size (h) reduced, the improvement in accuracy is called h-convergence. If the polynomial order (p) of the elements is increased, the response is called p-convergence. Analysts are encouraged to try increasing the model resolution to see if the results have converged. If a finer mesh significantly changes the response, the original model had not reached convergence, and there is no guarantee that the finer model has either.

1.3.3 Structural analysis equations

NOTATION USED:

K = stiffness matrix	M = mass matrix	C = damping matrix
U = displacements	U′ = velocity	U″ = acceleration
P = applied load	P_{cr} = buckling load	K_s = stress stiffening
Φ = mode shape	ω = forcing frequency	ω_n = natural frequency

Linear static analysis: small displacement, linear material. Solve by a variety of techniques such as Gauss elimination or Cholesky decomposition.

$$K\,U = P \tag{1.26}$$

Nonlinear static analysis: contact, plasticity, large displacements. Solve by some form of a Newton's method or other iterative scheme.

$$K(U)\,U = P(U) \tag{1.27}$$

Linear buckling analysis: eigenvalue problem. Solve by Lanczos's method.

$$[K + \lambda_{cr}\,K_s\,]\,\Phi = 0,$$

where

$$P_{cr} = \lambda_{cr}\,P$$

and

$$K_s = K_s(P) \tag{1.28}$$

Linear transient analysis: general time varying load. Solve by a numerical integration technique

$$M\,U'' + C\,U' + K\,U = P(t). \tag{1.29}$$

Nonlinear transient analysis: general time-varying load with contact, nonlinear materials, and large displacements. Typically, solve by an implicit integration for mildly nonlinear, and explicit integration for highly nonlinear:

$$M\,U'' + C\,U' + K(U)\,U = P(t,U) \tag{1.30}$$

Direct-frequency response analysis: steady-state harmonic condition. Solve like a linear static analysis except with complex mathematics.

$$P = \mathit{P}\,e^{i\omega t} => U = \mathit{U}\,e^{i\omega t}$$

$$[-\omega^2\,M + i\omega\,C + K]\,\mathit{U} = \mathit{P} \tag{1.31}$$

Real-natural-frequency analysis: no damping, *no* load. Solve with an eigenvalue technique, such as Lanczos method.

$$[-\omega_n^{\,2}\,M + K]\,\Phi = 0 \tag{1.32}$$

Modal-frequency response analysis: approach creates uncoupled equations. The substitution of $U = \Sigma\ z_j\ \Phi_j$ creates diagonal coefficient matrices $k = \Phi^T K \Phi$ and $m = \Phi^T M \Phi$, which reduces the direct frequency response equations to:

$$[-\omega^2\ m + i\omega\ c + k]\ z = \Phi^T P \tag{1.33}$$

1.3.4 Thermal analysis with finite elements

Heat transfer problems are commonly solved by finite difference or finite element methods. In the finite element approach, the temperature is assumed to vary over an element according to a simple polynomial relationship as shown by the shape functions

$$T(x) = N_1\ T_1 + N_2\ T_2 = \Sigma\ N_j\ T_j = [N]\{T\},$$

$$N_1 = 1 - x/L\ ,$$

and

$$N_2 = x/L\ , \tag{1.34}$$

where N_1 and N_2 are the same as the structural shape functions in Fig. 1.8. The thermal gradient and thermal flux can be written as a function of nodal variables T:

$$dT/dx = (d/dx)\ \Sigma\ N_j\ T_j = \Sigma\ dN_j/dx\ T_j = [B]\ \{T\},$$
$$q = -\kappa\ dT/dx = [G]\ [B]\ \{T\}, \tag{1.35}$$

where κ is the material conductivity. From variational principles, the thermal conductivity matrix can be derived as

$$k = \int B^T G B dV\ . \tag{1.36}$$

For the simple 1D rod, the thermal conductivity and structural stiffness matrix can be compared:

$$\text{Thermal: } k = \frac{A\kappa}{L}\begin{bmatrix} 1 & -1 \\ -1 & 1 \end{bmatrix},$$

$$\text{Structural: } k = \frac{AE}{L}\begin{bmatrix} 1 & -1 \\ -1 & 1 \end{bmatrix}. \tag{1.37}$$

Thus, if E is replaced by κ, the structural element becomes a conducting element. By analogy, all common structural elements (1D, 2D and 3D) can become heat-conducting elements with a change of material properties. For an additional discussion of finite elements in heat transfer, see Refs. [3] and [4].

1.3.5 Thermal analysis equations

NOTATION USED:		
K = conduction matrix	**C = capacitance matrix**	**R = radiation matrix**
T =temperatures	**Q = applied flux**	**H = convection matrix**

Linear steady-state analysis: linear properties and constant convection coefficients. Solve by a linear solver.

$$K\,T + H\,T = Q \qquad (1.38)$$

Nonlinear steady-state analysis: radiation, temperature dependent properties. Solve by some form of Newton's method.

$$R\,T^4 + K(T)\,T + H(T)\,T = Q(T) \qquad (1.39)$$

Nonlinear transient analysis: general time varying loads and boundary conditions. Solve by numerical integration.

$$C\,T' + R\,T^4 + K(T)\,T + H(t,T)T = Q(t,T) \qquad (1.40)$$

Some issues involved with heat transfer analysis include

≺1≻ surface elements that are required for convection and radiation, and their associated convection coefficients and emissivities;

≺2≻ the radiation view factor matrix is very costly to compute; and

≺3≻ thermal and structural models that may use different meshes such that the nodal temperatures must be interpolated from the thermal model to the structural model.

1.4 Symmetry in FE models

1.4.1 General loads

There are techniques within FEA to take advantage of symmetry within a structure to reduce the model size and computer resources required. In its most general form, the structure and boundary conditions (BC) must be symmetric;

however, the loads may be general. In Fig. 1.11, an example with one plane of symmetry shows that a general load case can decompose into both a symmetric case and an antisymmetric case. In this example, only half of the structure is modeled.

This approach requires some effort from the analyst to calculate the symmetric load (Ps) and antisymmetric load (Pa). Some programs, such as MSC/Nastran have automated this technique in a cyclic symmetry solution.

1.4.2 Symmetric loads

A common special case is a symmetric structure with symmetric loads shown in Fig. 1.12. To solve this problem, the model is one-half of the full structure with symmetric boundary conditions on the cut. Only loads appearing on the modeled half are used, with loads on the symmetry plane cut in half.

Typical models of mirrors on three-point supports are shown in Fig. 1.13. For natural frequencies or lateral g loads, a half model may be used. For axial g loads and isothermal temperature changes, a one-half or a one-sixth model may be

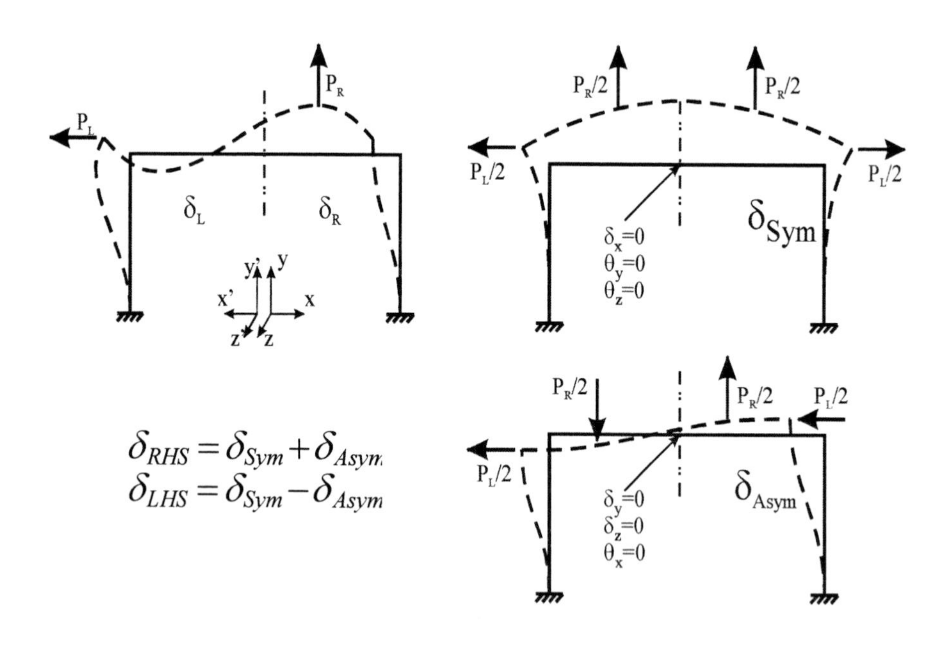

Figure 1.11 Symmetric structure with general load.

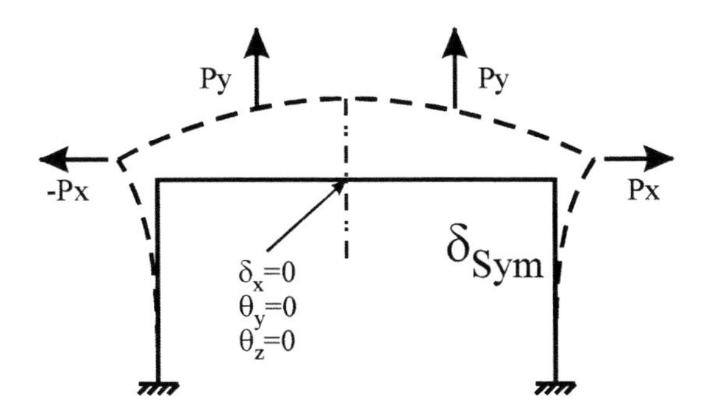

Figure 1.12 Symmetric structure with symmetric load.

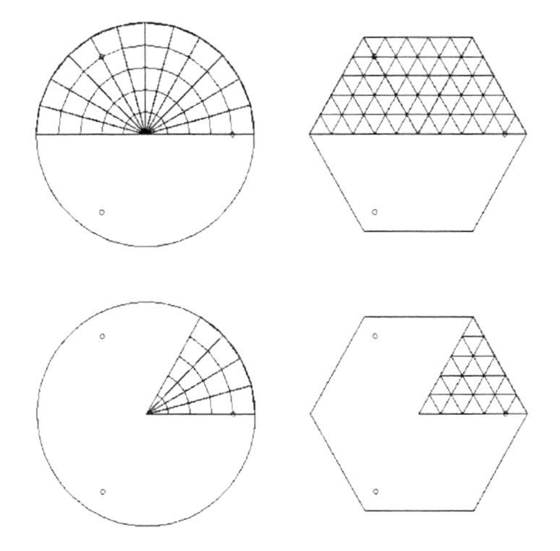

Figure 1.13 Typical symmetric models in optics.

used. Note that symmetric structures have antisymmetric mode shapes. Thus, when using a half model to calculate dynamic modes, modes must be calculated with symmetric BC, and again with antisymmetric BC, to find all modes. Both BC must be used in buckling analyses, since the lowest mode may be either symmetric or antisymmetric.

1.4.3 Modeling techniques

NOTATION USED:
d_N = displacement normal to the symmetry plane
d_{T1}, d_{T2} = displacements in the symmetry plane
Θ_N = rotation normal to the symmetry plane
Θ_{T1}, Θ_{T2} = rotations in the symmetry plane

For any general orientation, the model must use a displacement-coordinate system that aligns with the plane of symmetry. The appropriate boundary conditions are

Symmetric BC:	Antisymmetric BC	
$d_N = 0$	$\Theta_N = 0$	(1.41)
$\Theta_{T1}, \Theta_{T2} = 0$	$d_{T1}, d_{T2} = 0$	

For a pie-shaped sector of an optic (Fig. 1.13), cylindrical coordinates are naturally defined, so the symmetric BC are

$$d_\Theta = 0,$$

and

$$\Theta_R = \Theta_Z = 0. \tag{1.42}$$

If a node exists on axis of the cylindrical system, only axial displacement (d_Z) is free to move. For axial points, the symmetric BC are

$$d_R = d_\Theta = 0, \Theta_R = \Theta_Z = \Theta_Z = 0. \tag{1.43}$$

Solid elements are 3D in geometry, so they cannot lie in a plane of symmetry. However, 1D beams and 2D plates may lie in the plane of symmetry. An element lying *in* the plane of symmetry must have its stiffness cut by half so that upon reflection, the other half will be added. This action is not always the same as cutting a thickness by half and then computing the stiffness, since bending properties are a function of thickness cubed.

FOR 2D PLATES/SHELLS WITH ORIGINAL THICKNESS AT T_0:
Membrane thickness: $T_m = T_0/2$
Bending Inertia: $I_b = I_0/2 = 4(T_m^3/12)$
or bending ratio: $R_b = 4.0$
Transverse shear factor scales T_m, so use original R_s
Stress recovery: $z = T_0/2$, not $T_m/2$
FOR 1D BEAMS WITH ORIGINAL PROPERTIES A_0, I_0, J_0, K_0, C_0:
Cross-sectional area: $A = A_0/2$ (cut in half)
Bending inertia: $I = I_0/2$ (cut in half)
Torsional factor: $J = J_0/2$ (cut in half)
Transverse shear factor: $K = K_0$ (use original)
Stress recovery location: $C = C_0$ (use original)

1.4.4 Axisymmetry

Many optical elements, such as circular lenses, are axisymmetric in geometry and mounted in a ring-type mount. Another example would be a lightweight mirror sitting on an air-bag test support with the core structure modeled with smearable (effective) properties. For these applications, an axisymmetric model is an option if the loads are also axisymmetric such as axial g loads or axisymmetric temperature distributions.

The analyst may chose between using special purpose axisymmetric elements or creating a thin wedge with conventional 3D elements and symmetric BC (Fig. 1.14). The wedge model has the advantage that it can easily be expanded to a full 3D model to study nonsymmetric effects.

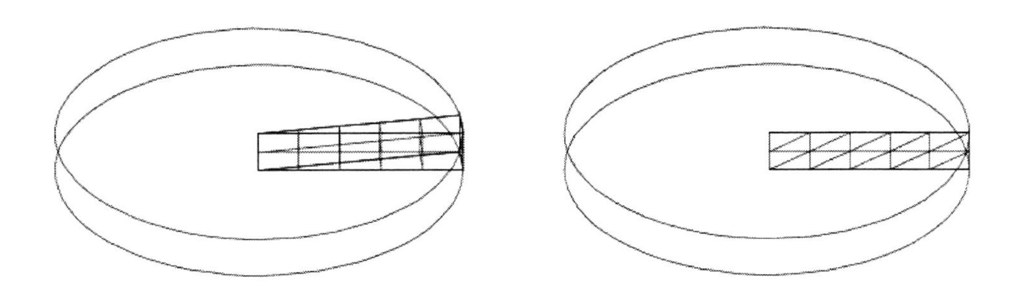

Figure 1.14 Axisymmetric models.

1.4.5 Symmetry: pros and cons

Even though a structure displays symmetry, the analyst must consider the pros and cons of using symmetric models before starting an analysis. A few of the considerations are as seen in Table 1.2.

Table 1.2 Considerations for the use of symmetric models.

ADVANTAGES OF USING SYMMETRIC MODELS:

<1> Faster modeling.
<2> Faster run times.
<3> Smaller input and output files.
<4> Smaller databases.

DISADVANTAGES OF USING SYMMETRIC MODELS:

<1> Requires multiple solutions and combinations if it is an asymmetric load.
<2> Requires multiple BC runs to get all dynamic and buckling modes.
<3> Requires interpretation of results for imaged segments.
<4> Cannot get full model plots easily.

1.5 Stress Analysis

1.5.1 Stress models

Stress prediction, requiring the derivatives of displacement (strains), is one order lower in accuracy than displacement prediction. Any FE model predicts displacements more accurately than it predicts stress. In other words, good stress models require more detail than displacement or dynamics models. The extra detail is required where stress gradients are high, such as around mounts. Common modeling techniques available for stress analysis include detailed models and coarse models.

In detailed models, all significant geometric detail of stress risers is included (e.g., fillets, holes etc.). The high detail is very time consuming; many require mesh convergence studies to gain confidence in the predictions.

If the component of interest is represented by a coarse model, and the net cross-section forces are accurate, the nominal (average) stresses may be accurate, but the results around stress risers are not accurate. For these models, the analyst may estimate peak stress using the cross-sectional forces. For some geometries, the nominal stresses may be multiplied by handbook[6] values for stress concentration factors (K_t) to estimate peak stress:

$$\sigma_{peak} = K_t * \sigma_{nominal} \,. \tag{1.44}$$

Figure 1.15 shows a primary mirror bipod flexure mount modeled with coarse beam elements (24 beam elements), while Fig. 1.16 shows a highly detailed solid mesh (>100,000 solid elements). The nominal beam stresses are scaled by published values for K_t for the appropriate element load. When the scaled stresses are compared to the highly detailed mesh, they agree within 7%. Thus, the proper use of engineering judgement can save significant analysis resources.

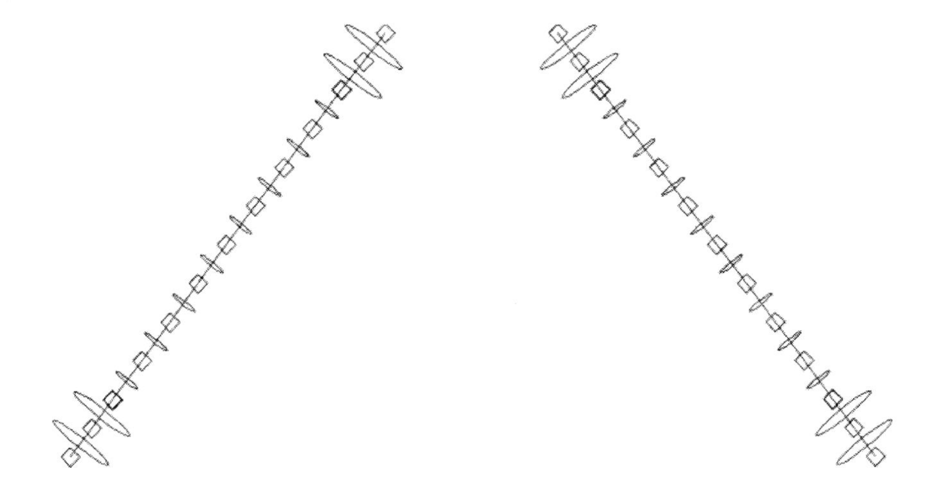

Figure 1.15 Coarse beam model of bipod flexure.

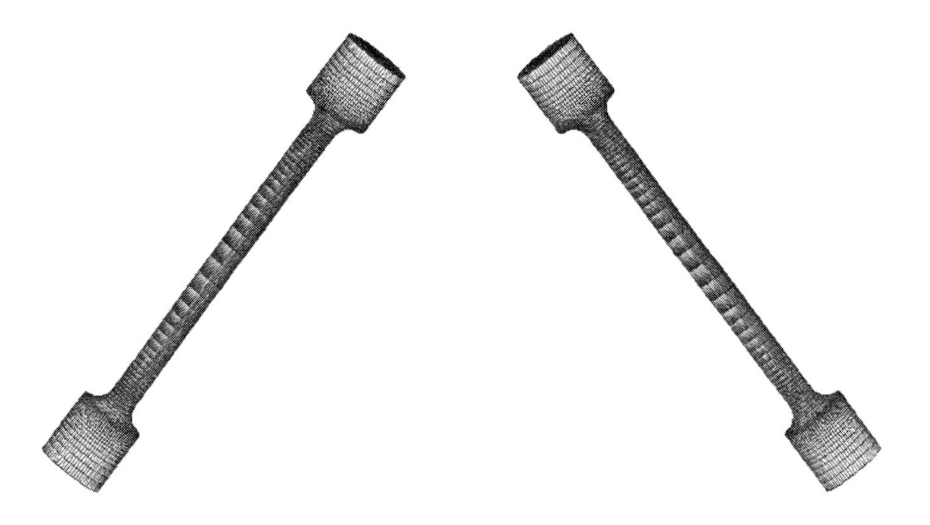

Figure 1.16 Detailed solid model of bipod flexure.

1.5.2 Failure theories

For ductile materials (most metals), a common failure criterion in optical structures is yielding, since permanent deformation causes misalignment. In typical mechanical structures, a 0.2% offset permanent strain condition is used. Whereas in many optical metering structures, the criterion of microyield (1.e–06 offset strain) is used. The maximum distortion energy criterion predicts yielding when $\sigma_{vm} > \sigma_y$.

In brittle materials (glasses, ceramics), the material fails by fracture. This ultimate condition occurs when the stress intensity exceeds fracture toughness. From fracture mechanics, the propagation of flaws or cracks is the mechanism for failure of homogeneous materials. The resistance to flaw growth is fracture toughness, which is a material property. In optics, the grinding process creates sizeable flaws on the surface, which the polishing process attempts to remove. Even when an inspection detects none, flaws exist which are smaller than the detectable limit.

To analyze for fracture, compute the stress intensity factor, K_I, K_{II}, K_{III}, for the mode of fracture (Fig. 1.17) and compare it to the fracture toughness, K_{IC}, K_{IIC}, K_{IIIC} of the material. If $K > K_c$, then the crack will propagate. Within FEA, there are several techniques to compute a fracture failure. Options 1 and 2 above, which involve detailed modeling of individual cracks, are typically applied when "significant" cracks have been observed in an optic. For nominal conditions, Options 3 and 4 use standard modeling techniques.

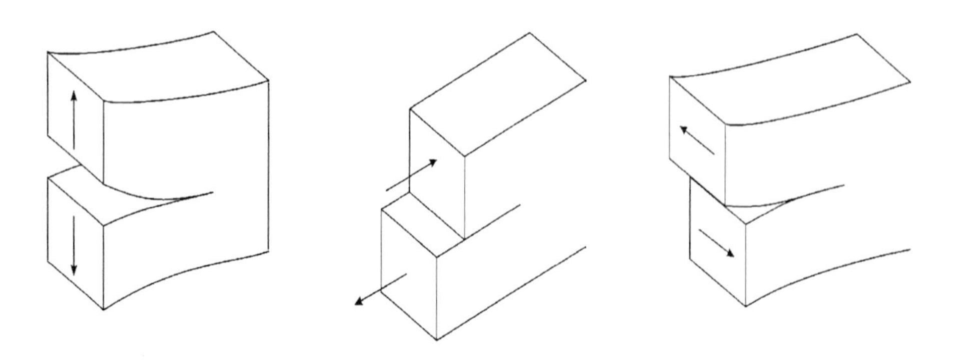

Figure 1.17 Modes of fracture.

OPTION 1: ***Detailed model using a crack tip element.***

Some FEA codes offer a specialized element that reports stress intensities, K_I, K_{II}, K_{III} of a modeled crack. The CRAC2D and CRAC3D elements within MSC/NASTRAN are crack-tip elements. These require detailed modeling of individual cracks. These may be used for mixed mode problems .

OPTION 2: ***Detailed model using a strain-energy release rate.***

A technique available to all FEA programs is to model an individual crack as a "slit" of unequivalenced nodes in a fine mesh. A first run calculates strain energy, SE_1. The model is then edited to extend the crack a small amount, giving a change in crack surface area of ΔA. The analysis is rerun to find strain energy, SE_2. The strain energy release rate, G, and stress-intensity factor K_I can be found from

$$G = \frac{(SE_2 - SE_1)}{\Delta A}$$
$$K_I = \sqrt{EG} \qquad (1.45)$$

OPTION 3: ***Coarse model without modeling the crack.***

This technique is useful for small cracks distributed in the component. Run an analysis on the coarse component model to find maximum principal stress, σ_1, for mode I. If the load is reversible, also find the minimum principal stress, σ_2. Estimate the size of a crack in the component. Then use a stress intensity factor from a handbook [7, 8] equation as in Fig. 1.18.

OPTION 4: ***Compare stresses to test samples.***

Run experimental tests on representative samples to find breaking strength. Model the test samples with FE to find nominal stress at failure, σ_f. When modeling the prime hardware use the same element size as in the test sample model at critical joints. Compare the stress from prime hardware model to σ_f to predict failure. In this approach, Weibull statistics[9] are needed to obtain σ_f to account for flaw distribution and part size.

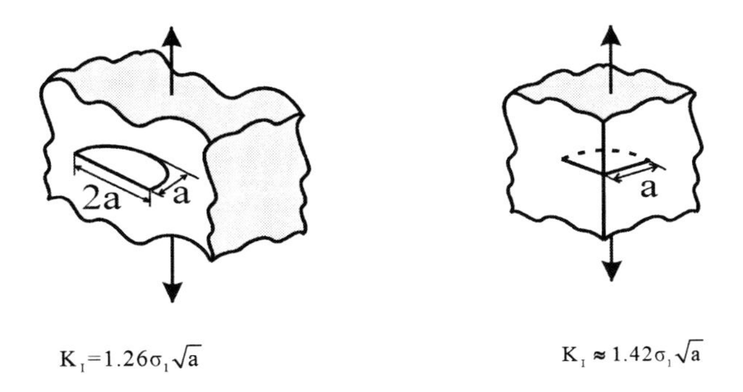

$K_I = 1.26\sigma_1\sqrt{a}$ $\qquad$ $K_I \approx 1.42\sigma_1\sqrt{a}$

Figure 1.18 Stress intensity factors.

1.5.3 Stress plots

During the post-processing phase of an analysis, there are many options for plotting stress. For ductile materials, Von Mises stress should be plotted for comparison to yield or micro-yield. Whereas for brittle materials, maximum principal stress should be plotted for failure analysis as discussed in the previous section. If the load is reversible, so is the stress, making a plot of minimum principal stress appropriate for reversed loading. Directional stress plots are useful to understand the behavior of the structure so that design improvements can be made.

Before making any stress plots, the analyst must understand several issues or the plots are completely meaningless. In most codes, the analysis calculates stress in each element's individual coordinate system. These stresses may be calculated at centroids, corners, or integration points. The post-processor can then plot these stresses in element coordinate systems, local systems, or global systems. The analyst should understand the averaging domains so that high element stresses are not overlooked when averaged with many lower stresses. Also, note that, unlike directional stresses, Von Mises and principal stresses cannot be averaged between elements. Von Mises and principal stress must be recomputed from averaged directional stresses. The analyst should run several simple test cases to fully understand all post-processing options. When reporting stresses, the type of stress plotted, the options used, and the load case description should be noted on the plot.

1.5.4 Load envelopes

In aerospace applications, launch loads are typically given as an envelope of maximum scale factors in orthogonal directions. For a single launch event, typical accelerations may given as

±8g along X,	±4g along Y,	±6g along Z,
±2rad/s^2 about X,	±2rad/s^2 about Y, and	±4rad/s^2 about Z,

for a combination of 2^6, or 64, load combinations. If temperature extremes are also considered, there are 128 combinations. Another launch event would create an additional 128 combinations.

An efficient analysis technique is to run seven unit-load cases of 1g along each axis, 1 rad/s^2 about each axis, and an 1° isothermal change. A post-processing program[10] is then used to step through all combinations to scale stresses, combine, recalculate Von Mises and principal stresses, and store the maximum values. A single plot of these maximums then represents a "high-water" or envelope of peak stresses, greatly reducing the post-processing effort.

1.6 Vibrations

1.6.1 Natural frequencies

The first step in any vibration analysis is to calculate the natural frequencies of the structure. For lightly damped structures, typical of stiff optical systems, damping has little effect on the natural frequency and can be ignored. Free, undamped vibration can be calculated from:

$$M\,U'' + K\,U = 0. \tag{1.46}$$

For harmonic motion:

$$U = \Phi\, e^{i\omega t},$$

and

$$U'' = -\omega^2\, \Phi\, e^{i\omega t}, \tag{1.47}$$

Upon substitution, an eigenvalue problem is obtained,

$$(-\omega_{nj}^2 M + K)\, \Phi_j = 0, \tag{1.48}$$

where the eigenvector (Φ_j) is the mode shape for j^{th} vibration mode at a natural frequency of ω_{nj} radians/sec, or f_{nj} Hz (cycles/sec)

$$\omega_{nj} = 2\,\pi\, f_{nj}. \tag{1.49}$$

Natural frequencies are important physically because resonances occur at natural frequencies in the form of the mode shape. Since no load has been applied in a natural frequency analysis, the magnitude of the response is meaningless. However, the shape (or relative values) of displacement, stress, or strain energy provides insight into possible design improvements.

Natural frequencies are important mathematically because the mode shapes (Φ) are principal coordinates of the system and can be used to uncouple the system equations in the modal analysis approach in the next section.

1.6.2 Harmonic response

Harmonic response analysis (or frequency response analysis) is the steady-state response to a harmonic forcing function. For a forcing function $P = \mathit{P}\, e^{i\omega t}$ of magnitude P at a forcing frequency ω, the displacement at a steady state is $U = \mathit{U}\, e^{i\omega t}$ with a response amplitude of U at the forcing frequency ω. The response amplitude can be determined from

$$[-\omega^2 M + i\omega C + K\,]\, \mathit{U} = \mathit{P}. \tag{1.50}$$

For each value of ω, a coupled system must be solved using complex mathematics.

The natural frequency mode shapes can be used to create uncoupled vibration equations in the modal approach. Since the eigenvectors are principal coordinates of the system, any response can be created by a linear combination:

$$U = \Sigma\, z_j\, \Phi_j\,. \tag{1.51}$$

From the orthogonality condition of eigenvectors,

$$k = \Phi^T K \Phi\,,$$

and

$$m = \Phi^T M \Phi, \tag{1.52}$$

the matrices k and m are diagonal. Upon substitution in the harmonic response equation, an uncoupled (diagonal coefficient matrices) system results,

$$[-\omega^2\, m + i\omega\, c + k\,]\, z = \Phi^T P = p\,, \tag{1.53}$$

if modal damping (c) is assumed. The solution of an uncoupled system may be written directly as

$$z_j = p_j \,/\, [(k_j\text{-}m_j\omega^2) + ic_j\omega]\,, \tag{1.54}$$

where z_j is participation of mode j at the current forcing frequency. Note that z is a complex number due to damping. Any physical response (U,σ) is then the combination of modal responses (Φ,S):

$$U = \Sigma\, z_j\, \Phi_j\,,$$

and

$$\sigma = \Sigma\, z_j\, S_j\,. \tag{1.55}$$

This feature is used in Chapter 8 to decompose optical surface motion into surface pointing and surface distortion components. Modal analysis is exact if all modes are used, but is approximate if the mode list is truncated to the lowest n modes.

A transfer function (TF) or frequency response function (FRF) is the steady-state response of any quantity tabulated over a frequency range of interest. These are often used to characterize a system dynamically, and are used in the random analysis described in the next section.

1.6.3 Random response

In many applications, the loading is only described by its statistical nature or load-power spectral density (PSD_P), which represents the energy content as a function of frequency. The statistical nature of output, or response-power spectral density (PSD_R), can be determined by random response analysis as given in Fig. 1.19, or the following:

$$PSD_R = FRF^2 \times PSD_P. \tag{1.56}$$

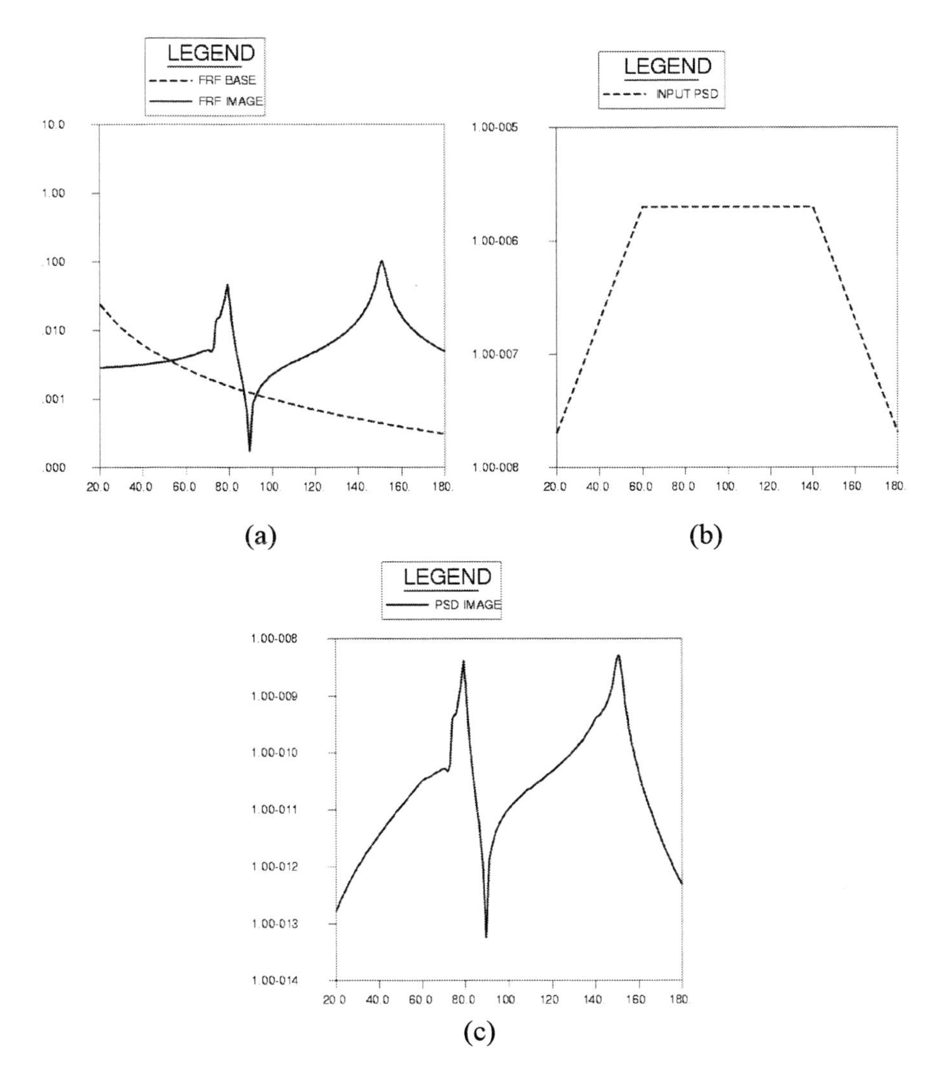

Figure 1.19 (a) Transfer Function of Image Motion (TF or FRF), (b) PSD_P of base shake input, and (c) PSD_R of image motion response.

If the area (A) under the response curve is determined from integration, the RMS of the response quantity is found from:

$$\text{rms} = 1\sigma \text{ value} = \sqrt{A}\,. \tag{1.57}$$

For normally distributed (Gaussian) behavior, the peak response has the following statistical behavior:

1σ value = 1 × rms => peaks are less than 1σ for 68.3% of the time
3σ value = 3 × rms => peaks are less than 3σ for 99.7% of the time

For more detailed discussion of vibration analysis, see Refs. [11] and [12].

1.6.4 Damping

In the harmonic response equations above, viscous damping is represented by the term "C". The presence of damping causes the response throughout a structure to reach peak values at different times, which may be referred to as a delay (in time units) or as a phase angle (in degrees). Typical sources of damping include

<1> joint relative motion with friction or fretting,
<2> plasticity or viscoelastic behavior, and
<3> air flow damping effects.

As pointed out by Richard,[13] damping is not well understood and cannot be predicted from theory. Instead, damping must be determined from testing. For structures subjected to cyclic loading, the energy absorbed by hysteresis effects may be used to determine an equivalent viscous damping ratio.

Damping is often specified as damping ratio (ζ), "fraction of critical damping," or "percent of critical damping." For use in the dynamic response equations, convert percent values to fractional values:

$$\zeta = \text{damping ratio} = C/C_{cr}\,. \tag{1.58}$$

Damping may also be specified as loss factor(η) or quality factor (Q):

$$\eta = \text{loss factor (g in Nastran)},$$

$$Q = \text{quality factor (or amplification factor)}, \tag{1.59}$$

where

$$2\zeta = \eta = 1/Q. \tag{1.60}$$

Optical structures are usually very lightly stressed, and are often used in a vacuum, which combine to give damping ratios in the range of $\zeta = 0.001–0.020$ which are typically frequency dependent.

1.7 Model Checkout

The results of any FE analysis should be considered guilty until proven innocent. The analyst must take full advantage of the model pre-processor to conduct as many model checks as possible, including

≺1≻ duplicate nodes and elements,
≺2≻ free boundaries,
≺3≻ surface normals, and
≺4≻ element geometry quality.

As additional proof of a valid model, checkout runs should be made as follows.

RIGID-BODY ERROR CHECK:

≺1≻ Remove all real BC; ground one node in all six DOF.
≺2≻ Apply six load cases of unit motion in each DOF at the grounded node.
≺3≻ Motion should be exactly stress free.
≺4≻ This finds "hidden" reactions to ground, bad MPCs, and bad autospc.
≺5≻ Image-motion equations should satisfy this check.

FREE BODY MODES CHECK

≺1≻ Remove all boundary conditions.
≺2≻ Calculate natural frequencies.
≺3≻ Model should have six, and *only* six, zero modes (rigid body modes).
≺4≻ Compare values of f_{1-6} to f_7 as an indication of modeling problems:

$$f_{\text{rigid-body}} << f_{\text{elastic}}.$$

≺5≻ This finds hidden reactions to ground and bad MPCs.
≺6≻ This also finds mechanisms that prevent static solutions.

STATIC LOAD CHECK

≺1≻ Apply a 1g static load case in each coordinate direction.
≺2≻ Check that it is reasonable and compare to intuition.
≺3≻ Compare to known solutions of similar structures.
≺4≻ Check for expected symmetry in results.
≺5≻ Check equilibrium (ΣForces = ΣReactions).
≺6≻ Check error measures and warning messages.

<7> Check mass property summary; compare to known mass and cg.

THERMAL SOAK CHECK

<1> Temporarily set all CTE to a common value (e.g., 10 ppm).
<2> Apply a kinematic boundary condition (e.g., one node for all six DOF).
<3> Apply a significant isothermal load (e.g., 100°).
<4> Check to see if all resulting stresses are zero.
<5> This locates rigid links or constraints that prevent thermal growth.

In addition, the analyst should make as many checks as possible against hand solutions, classical textbook solutions, and engineering intuition. Whenever possible, compare results to experimental data from prototype structures or previous designs.

1.8 Summary

Modern analysis tools are very valuable, but the burden is still on the *engineer* to:

<1> Understand structural/thermal/optical theory.
<2> Understand FE theory and assumptions.
<3> Understand details of the analysis program.
<4> Understand details of the pre/post-processor program.
<5> Make modeling decisions and assumptions.
<6> Verify the model.
<7> Interpret the results and draw conclusions.
<8> Document the model and assumptions, and report the results.

References

1. MSC/NASTRAN and MSC/PATRAN is a product of the MacNeal-Schwendler Corporation, Los Angeles, CA.
2. Knight, C., *The Finite Element Method in Mechanical Design*, PWS, Boston (1993).
3. Logan, D., *A First Course in the Finite Element Method, 3rd Ed,*. Brooks/Cole, Pacific Grove, CA (2002).
4. Cook, Malkus, Plesha, Witt, *Concepts and Applications of Finite Element Analysis, 4th Ed..*, Wiley, New York (2002).
5. MacNeal, R., *Finite Elements: Their Design and Performance*, Marcel-Dekker, New York (1994).
6. Young, W. C., *Roark's Formulas for Stress and Strain, Sixth Ed.*, McGraw Hill, New York (1989).
7. Broek, D., *Elementary Engineering Fracture Mechanics*, Noordhoff International Publishing, Leydon (1982).
8. Hertzberg, R., *Deformation and Fracture Mechanics of Engineering Materials, Fourth Edition*, Wiley, New York (1996).

9. Harris, D., *Infrared Window and Dome Materials*, SPIE Optical Engineering Press, Tutorial Texts Vol. TT10, Bellingham, WA (1992).
10. Genberg, V., Vianese, J., "Enveloping results of multiple load cases," *Proc. of MSC Americas User Conference* (1998).
11. Craig, R., *Structural Dynamics, An Introduction to Computer Methods*, Wiley, New York (1981).
12. Steinberg, D., *Vibration Analysis for Electronic Equipment, 2nd Ed.*, Wiley, New York (1988).
13. Richard, R. "Damping and Vibration Considerations for the Design of Optical Systems in a Lauch/Space Environment," *Proceedings of SPIE*, **1340** (1990).

≺Chapter 2≻ Optical Basics and Zernike Polynomials

The goal of this chapter is to discuss common optical performance metrics and the basics of image formation such that the mechanical engineer may relate their designs, concepts, and response quantities to the performance of the optical system. The second half of this chapter addresses the use of orthogonal polynomials, such as the Zernike polynomials, to describe optical surface data. These polynomials are useful to describe optical surface deformations and wavefront error due to temperature and mechanical stress.

2.1 Electromagnetic Basics

A characteristic of all imaging systems is their ability to modify and reshape the incident electromagnetic radiation into an image. Thus, understanding the propagation of light is fundamental to our goal. Ultimately, we are concerned with how the environment impacts this propagation, but that will be addressed in later chapters. Light may be defined as a transverse electromagnetic wave where the electric and magnetic fields vibrate or oscillate perpendicular to the direction of propagation. For our purposes, the easiest way to consider light propagation is in 1D, as illustrated in Fig. 2.1. The mathematical equation describing the electric field vector, E, is given by

$$E(z,\tau) = Ae^{i(w\tau - kz)}, \tag{2.1}$$

where the electric field is a function of both position, z, and time, τ. The amplitude of the wave is denoted by A and the phase is given by, $\omega\tau\text{-}kz$, where

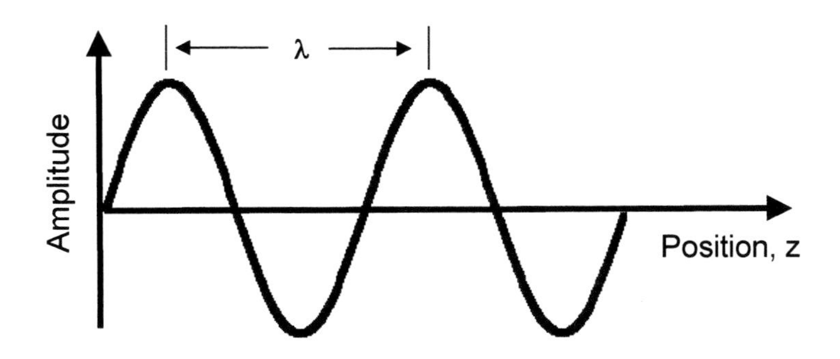

Figure 2.1 One-dimensional representation of an electromagnetic wave.

$k = 2\pi/\lambda$ is the wave number and ω is the angular frequency of the light wave. This representation may be extended to three-dimensional waves such as planar, cylindrical, and spherical waves typical of imaging systems.

When an electromagnetic wave enters a medium such as optical glass, the speed of the wave decreases. The ratio of the speed of the wave in a vacuum to the speed in a medium is called the *index of refraction*, *n*. The index of refraction for common optical glasses in the visible spectrum is roughly 1.5 to 2.0. The index of refraction for common IR materials ranges from 1.5 to 4.0.

The *wavelength*, λ, is the distance an electromagnetic wave travels in one cycle. Wavelengths in the visible spectrum range from 0.45 to 0.70 μm. Electromagnetic radiation may also be described by its *optical frequency*, ν, given in number of cycles per second (Hz). For example, the optical frequency for light at a wavelength of 546 nm is 5.5×10^{14} Hz. The relationship between wavelength and frequency is given by

$$\nu = \frac{c}{\lambda}. \tag{2.2}$$

Two electromagnetic waves are considered in-phase when the peaks and troughs for each wave coincide. Two waves out-of-phase with each other are shown in Fig. 2.2. Since a full cycle represents 360 deg or a wavelength, the phase difference between two waves may be expressed either in deg or in waves. For example, two waves out-of-phase by 90 deg are out-of-phase by a quarter wavelength. Two waves that have a phase difference that is an integer number of waves are considered in-phase since they perfectly overlap.

2.2 Polarization

Many optical systems use polarized light or polarizing optics. A well-known example is polarized sunglasses, which are often used to reduce the glare from water. The purpose of this section is to define and describe various states of polarization. In Chapter 4, we discuss how mechanical stress in transmissive optical materials affects the state of polarization.

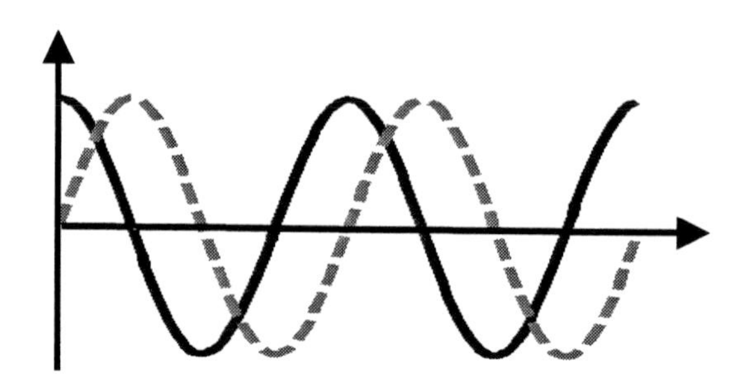

Figure 2.2 Two electromagnetic waves out-of-phase.

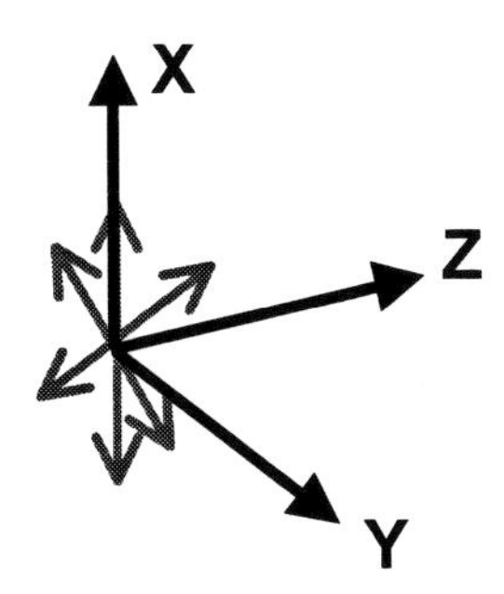

Figure 2.3 Unpolarized light (arrows represent the direction of the electric field).

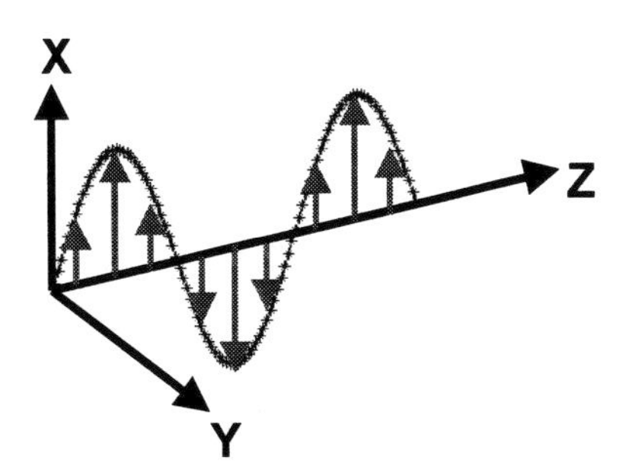

Figure 2.4 Linear polarization.

As stated earlier, light is a transverse electromagnetic wave where the electric field vibrates perpendicular to the direction of propagation. Light, where the direction of the electric field vector varies rapidly and randomly, is known as unpolarized light and is illustrated in Fig. 2.3. Natural light, for example, is unpolarized and the direction of the electric field changes approximately every 10^{-8} seconds. Linearly polarized or plane polarized light describes light whose electric field vector oscillates in a plane known as the plane of vibration as shown in Fig. 2.4. Here, the plane of vibration is the xz-plane, and the direction of the electric field moves up and down along the x-axis.

In general, since the electric field is a vector quantity, the electric field may be decomposed into components, E_x and E_y, along an arbitrary set of x and y-axes, respectively. The relative magnitude and phase of the components describes the state of polarization.

For linear polarization, the electric field components are in-phase with each other as shown in Fig. 2.5. In this example, the amplitudes of E_x and E_y are equal and their sum results in an electric field vector vibrating in a plane at 45 deg. Elliptical polarization occurs when E_x and E_y are out-of-phase. A special case of elliptical polarization is circular polarization which occurs when E_x and E_y are of equal amplitude and out-of-phase by 90 deg as shown in Fig. 2.6. Here the tip of the electric-field vector carves out a helix of circular cross section.

2.3 Rays, Wavefronts, and Wavefront Error

The propagation of light waves from a point source in an isotropic and homogeneous medium takes a spherical shape as shown in Fig. 2.7. At any instant in time, each surface joining all points of constant phase is called the wavefront. Neighboring surfaces of constant phase are separated by a wavelength. Rays, fictitious entities normal to each wavefront surface, are useful for understanding and analyzing optical systems. The optical distance traveled by a ray is known as the *optical path length*, or OPL. The OPL is computed as the physical distance a ray has traveled, s, multiplied by the index of refraction of the medium in which it travels as given by

$$\mathrm{OPL} = \int_a^b n(s)ds\,. \tag{2.3}$$

Across the surface of a given wavefront, the OPL is the same for each point. This is the basis for how images are formed by an optical system. For example, consider a diverging spherical wavefront incident upon a lens element shown in

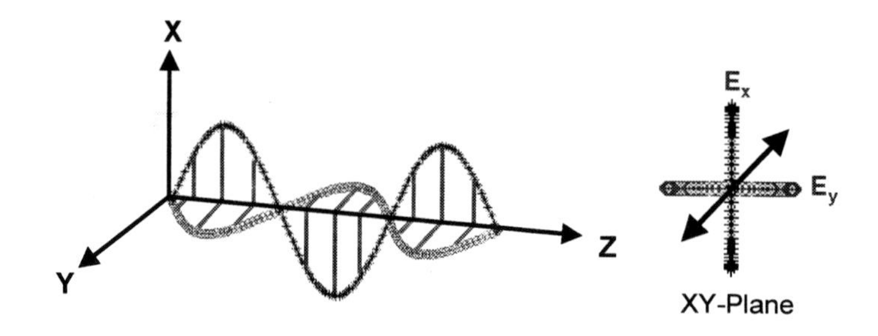

Figure 2.5 Linear polarized light at 45 deg.

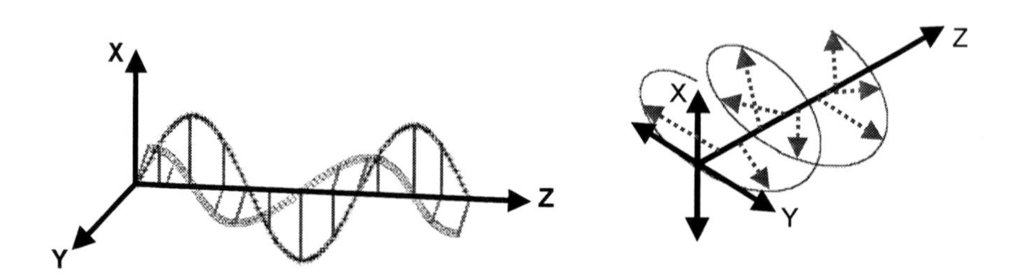

Figure 2.6 Circular polarization.

Fig. 2.7. After the wavefront passes through the lens element, the wavefront is converging. The reversal in the wavefront curvature is a consequence of the center rays traveling a greater distance through the lens element and slowing down relative to the edge rays.

For an optical system to form a perfect image point, the exiting wavefront must be spherical, and the rays normal to the wavefront must converge to the wavefronts' center of curvature. The departure of the OPL of the actual wavefront to a perfectly spherical reference wavefront measured over the wavefront surface is a measure of *wavefront error*. The difference in OPL is known as the *optical path difference* (OPD). An optical system producing a spherically aberrated wavefront, a common optical aberration, is shown in Fig. 2.8. Here the edge rays focus to a different point than the rays closer to the optical axis. Wavefront error may be quantified by the peak-to-valley error (p-v) and by the root-mean-square (rms) error. P-V errors represent the difference

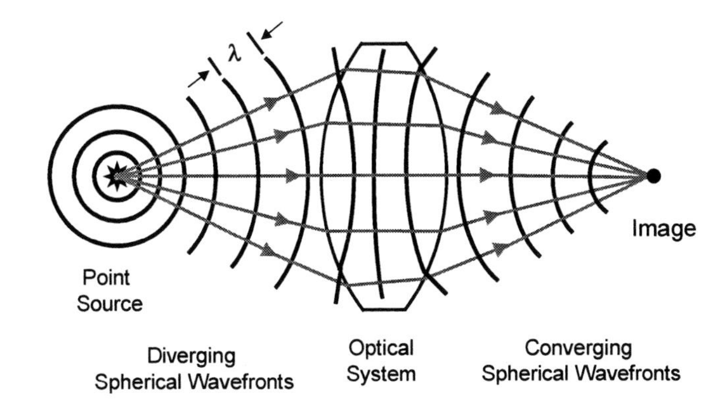

Figure 2.7 Lens element forming an image.

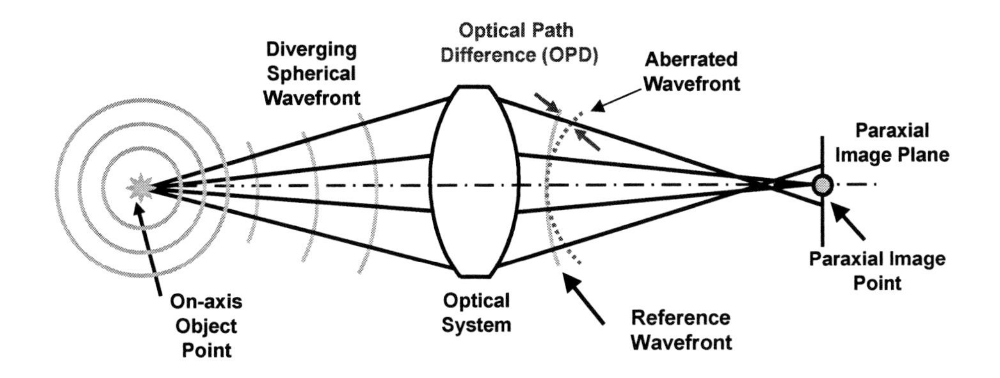

Figure 2.8 Lens element introducing wavefront error (spherical aberration).

between the maximum and minimum OPD over the wavefront. The rms is often a more meaningful measure of wavefront error since it is an average value over the wavefront surface. It is a simple extension to see how mechanical loads that deform the surface of an optical element create wavefront error. Furthermore, temperature and mechanical stress acting on an optical element also introduce wavefront errors by changing the index of refraction of the material.

2.4 Image Quality and Optical Performance

This section is designed to familiarize the mechanical engineer with various optical performance metrics. The first thing to understand regarding optical performance is that an optical system can never produce an image that is an exact duplicate of the object. A simpler way of considering this is that the image of a point object is never a point. The actual image formed by an optical system of a point source is a smeared or blurred point whose physical extent is referred to as the image blur, blur radius, or blur diameter. We will use the term blur diameter to describe the size of an image point. The goal for many imaging systems in optimizing optical performance is to minimize the size of the blur diameter. There are many factors that may contribute to the inability of an optical system to produce a perfect point, including the effects of diffraction, geometrical aberrations, fabrication errors, alignment errors, and, of course, environmental effects. However, if all is perfect, diffraction limits the quality of the image and hence provides the reference for which image quality is measured.

2.4.1 Diffraction

Diffraction is due to the wave nature of light and occurs at the boundary of obstacles in the light path that alter the amplitude and phase of an incident wavefront. The obstacle may be an aperture of an optical element or a mechanical support structure that causes the light to bend, or be redirected, from the paths predicted by geometrical optics. For example, the image of a point source at infinity for an optical system with a circular lens element is shown in Fig. 2.9. The interaction of the incident plane wavefront with the boundaries of the aperture results in the constructive and destructive interference of the exiting wavefront. The image produced by the focusing lens is not a perfect point, but a series of concentric light and dark rings. For an aberration-free system, the central bright spot is known as the *Airy disk* and contains 84% of the energy. The diameter of the Airy disk represents the smallest blur diameter that an optical system can produce and is given by

$$D = 2.44\lambda(f/\#), \tag{2.4}$$

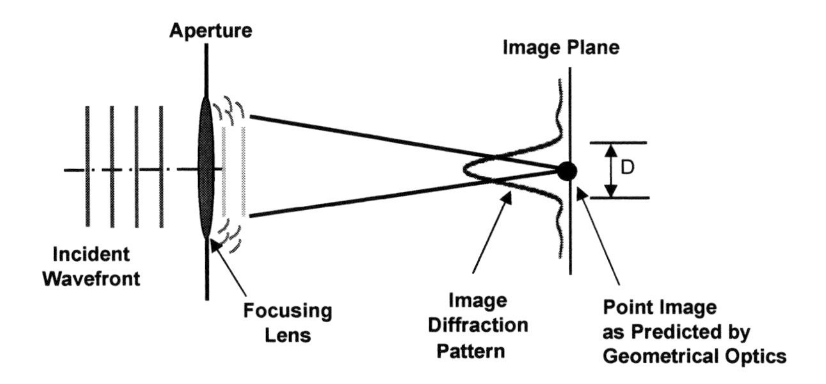

Figure 2.9 Diffraction image formed by a circular aperture.

where $f/\#$ (f-number) is a measure of the light collecting properties of an optical system. As a rule-of-thumb, for visible systems operating at a wavelength near 0.5 μm, the size of the Airy Disk is equal to the f-number in microns.

An optical system is known as *diffraction-limited* when the effects of diffraction dictate the size of the blur diameter. An acceptable amount of wavefront error may exist in an optical system where the system is still considered diffraction-limited. The allowable wavefront error is given by the Rayleigh criterion, which states that diffraction-limited performance is maintained for up to a quarter-wave of OPD p-v. This corresponds to an rms wavefront error of $\lambda/14$.

2.4.2 Measures of image blur

There are several optical performance metrics used to measure the size of the image blur or blur diameter. The type of performance metric used depends on the purpose of the optical system. When the blur diameter approximates the size of the Airy disk, which is typical for high-performance systems, then diffraction-based metrics must be employed. If the blur diameter is much larger than the Airy disk, the effects of diffraction may be ignored and geometric-based metrics may be used.

2.4.2.1 Spot diagrams

Spot diagrams are created by tracing a grid of rays from a single object point through an optical system and plotting their intersection with the image plane. Spot diagrams are geometric based and exclude the effects of diffraction. The distribution of points on the image plane is a measure of the size of the blur diameter. Commonly, an rms spot diameter is computed that encloses approximately 68% of the energy. Spot diagrams are useful to determine the aberrations present in an optical system since each aberration produces a characteristic pattern. A spot diagram of a singlet lens exhibiting spherical aberration is shown in Fig. 2.10.

2.4.2.2 Point spread function

The *point spread function* (PSF) is another measure of the size and shape of the image of a point source. The PSF calculation includes both the effects of diffraction and geometrical aberrations. The PSF for an aberration-free system and for an optical system with coma error is shown in Fig. 2.11 and Fig. 2.12, respectively. Both 3D isometric views and intensity plots of the PSF are shown. The intensity plot uses a logarithmic scale to reveal the ring structure of the PSF more clearly. Notice how the energy in the aberrated case is spread over a much larger diameter than the aberration-free system.

2.4.2.3 Encircled energy function

The encircled energy is a plot of the energy contained in circles of increasing diameter versus the diameter of the circle. The circles are centered on the image centroid. An example of the encircled energy is plotted in Fig. 2.13 for the aberration-free PSF and the aberrated PSF shown in Figs. 2.11–2.12.

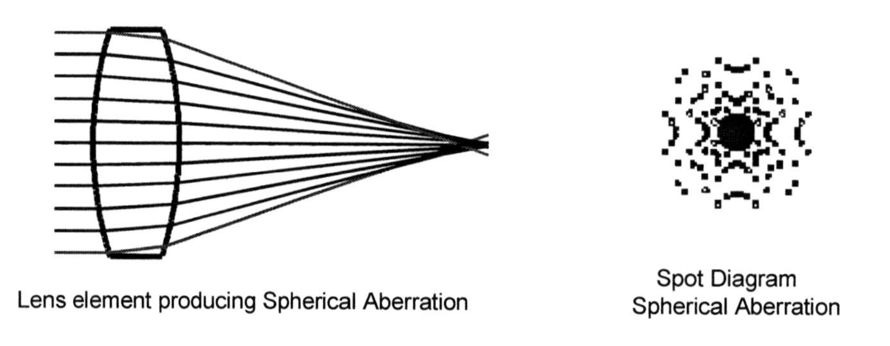

Figure 2.10 Spot diagram formed by a singlet lens exhibiting spherical aberration.

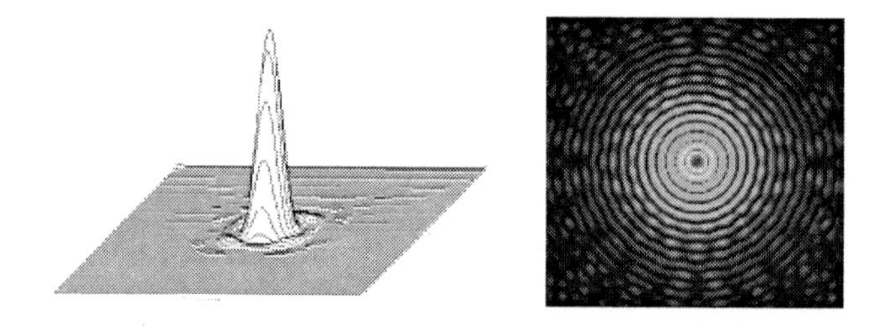

Figure 2.11 PSF for aberration-free system.

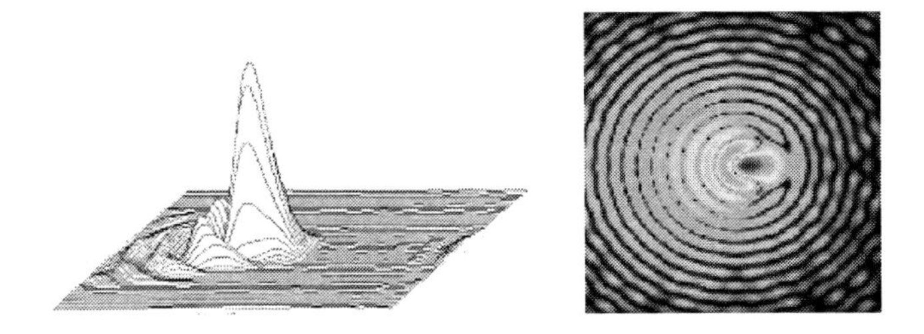

Figure 2.12 PSF for a system with coma error.

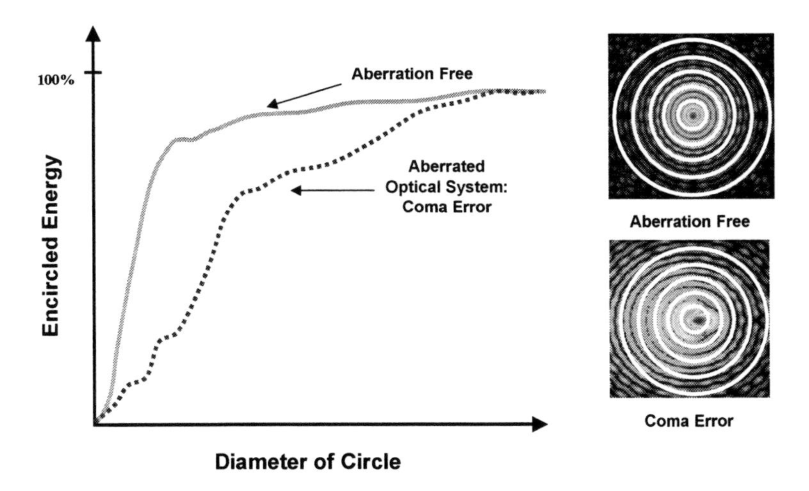

Figure 2.13 Encircled energy.

2.4.3 Optical resolution

The ability of an optical system to resolve two objects is a common measure of optical performance. For example, the Hubble Space Telescope can resolve two dimes from approximately 30 miles away. It should be clear that the effects of diffraction limit the resolution of an optical system as depicted in Fig. 2.14. As the diameters of the Airy disk for each image point increase, the intensity distributions begin to overlap and resolution decreases. Thus, the same parameters controlling the size of the Airy disk dictate the resolution of an optical system—namely, the system *f*-number and the wavelength of light. The combined diffraction pattern of the image of two points is shown as a function of *f*-number in Fig. 2.15. As the *f*-number of the optical system decreases, the combined intensity distribution begins to show two distinct peaks representing two object points. For example, the optical lithography industry increases

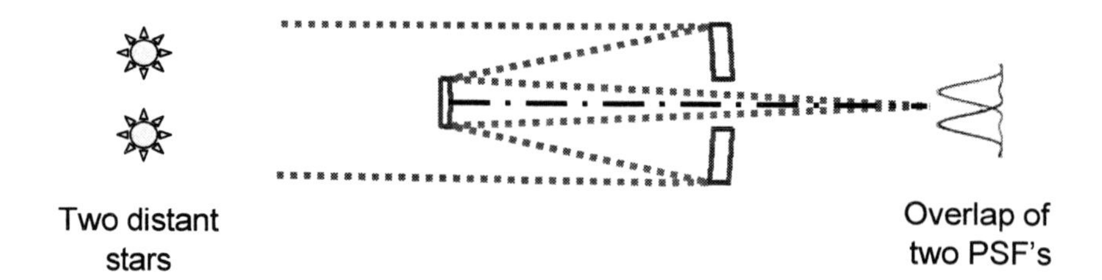

Figure 2.14 A measure of resolution is the ability of an optical system to resolve two point sources.

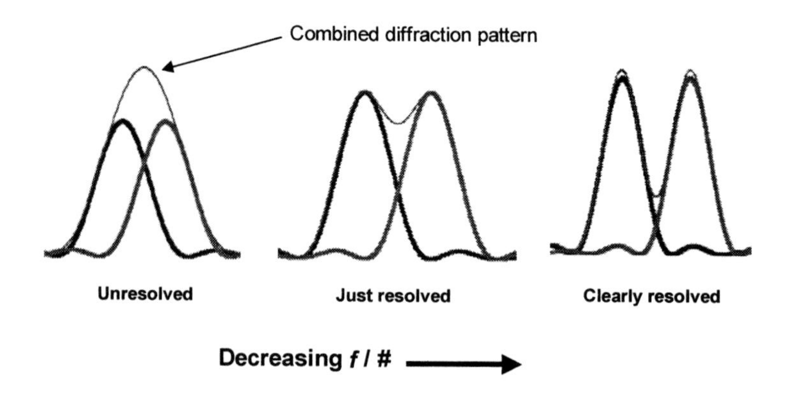

Figure 2.15 Combined intensity pattern for two point sources as a function of *f*-number.

resolution in their optical instruments by decreasing both the *f*-number and the wavelength of light. This increased resolution allows smaller feature sizes to be created on integrated circuits.

2.4.4 Modulation transfer function

A second, more comprehensive measure of the resolution of an optical system is given by the *modulation transfer function* (MTF). Here, the MTF considers the response of the optical system to sinusoidal intensity distributions of varying spatial frequency. This is illustrated for three spatial frequencies in Fig. 2.16. As the spatial frequency of each object is increased, the more difficult it is for the optical system to distinguish the peaks from the valleys. Resolving ability is quantified by the contrast ratio (also known as modulation), which is given by the following equation:

$$\text{Image Contrast} = \frac{I_{\max} - I_{\min}}{I_{\max} + I_{\min}}, \tag{2.5}$$

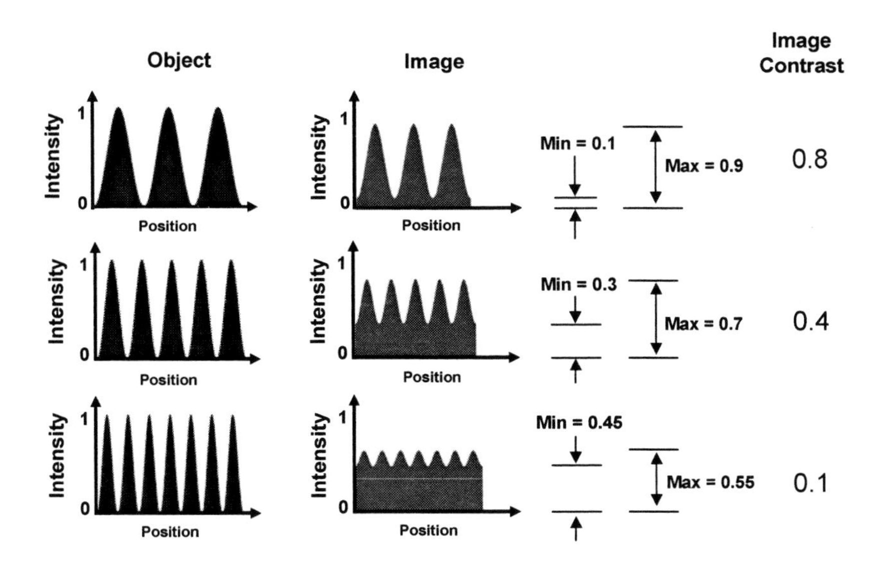

Figure 2.16 Image contrast computed for three spatial frequencies.

where I_{max} is the maximum intensity and I_{min} is the minimum intensity of the image. For the image to be an exact duplicate of the object the peaks would have a value of 1 and valleys would have a value of 0 yielding a contrast ratio of 1. As resolving capability diminishes, the contrast ratio decreases and there is little difference in the magnitude between the peaks and valleys. When the contrast ratio drops to 0, the optical system can no longer resolve the object, and a solid intensity pattern results. For incoherent light, the spatial frequency in which the optical system can no longer resolve is known as the cut-off frequency. The cut-off frequency, ρ_c, is a function of the *f*-number and is given as

$$\rho_c = \frac{1}{\lambda(f/\#)}. \tag{2.6}$$

The MTF curve is computed by plotting image contrast as a function of spatial frequency, and is shown for a diffraction-limited system and an aberrated system in Fig. 2.17. Notice how the diffraction-limited system is able to resolve higher spatial frequencies as compared to the aberrated system. The MTF is a valuable quantitative description to understand the resolving capability of an optical system by measuring image contrast over a range of spatial frequencies. Often the mid and low-end spatial frequencies are important to image quality, not just the cut-off frequency.

For linear systems, the total MTF for a system is equal to the product of the MTFs of each of the components that make up the system. The values at each spatial frequency are just multiplied together. For example, consider a photograph generated by a digital camera using a telephoto lens. The MTF of the

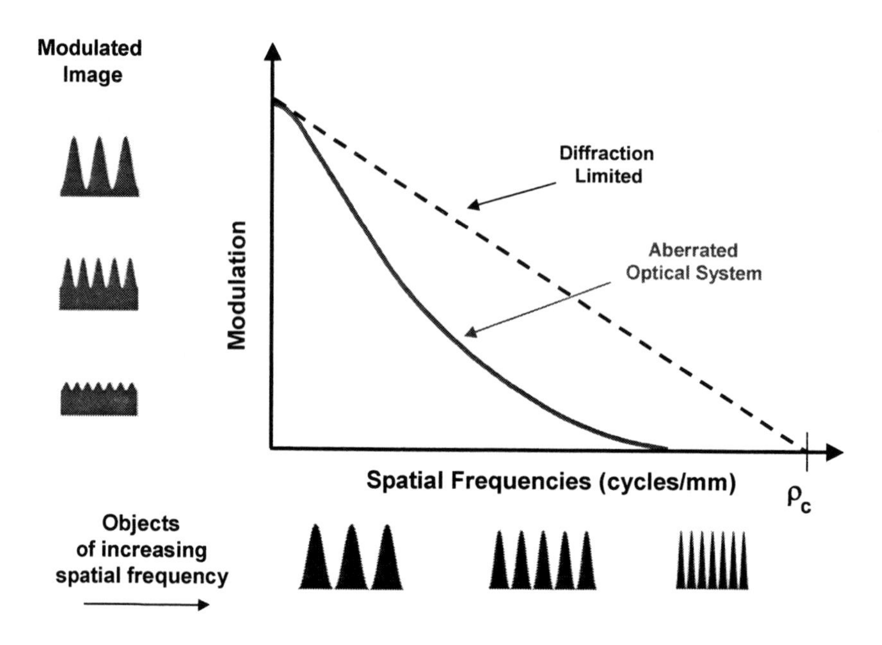

Figure 2.17 Modulation transfer function.

photo is the product of the MTF of the camera, the telephoto lens, and the detector array.

2.5 Image Formation

Linear systems theory may be used to describe a broad category of physical systems including many optical and mechanical systems. The response of these systems may be characterized by their impulse response and transfer functions. (Here, we will consider image formation only for spatially broad, incoherent light such as from incandescent light bulbs and the sun). Consider a single degree-of-freedom mechanical system as shown in Fig. 2.18. Subjecting this system to a unit impulse forcing function (an infinitesimally short duration impact force) produces a displacement of the mass known as the impulse response. The transfer function of the mechanical system is the Fourier transform of the impulse response. In analogous fashion, an optical system imaging a point source yields the impulse response of the optical system, which is simply the PSF as shown in Fig. 2.19. Taking the Fourier transform of the PSF yields the *optical transfer function* (OTF). Both the impulse response and the transfer function represent physical characteristics of the mechanical and optical system, and either may be used to compute the response of the system.

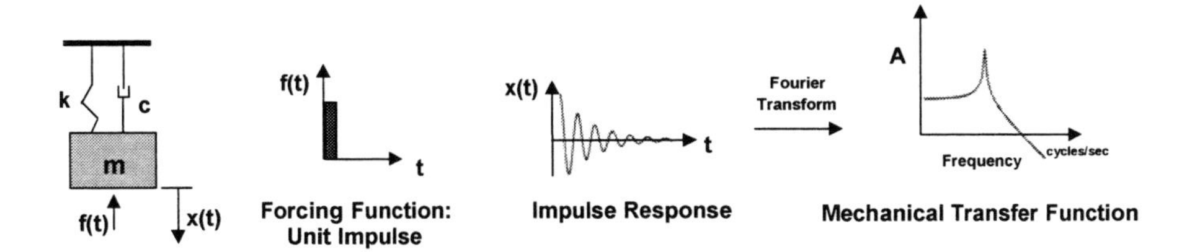

Figure 2.18 Mechanical system impulse response and transfer function.

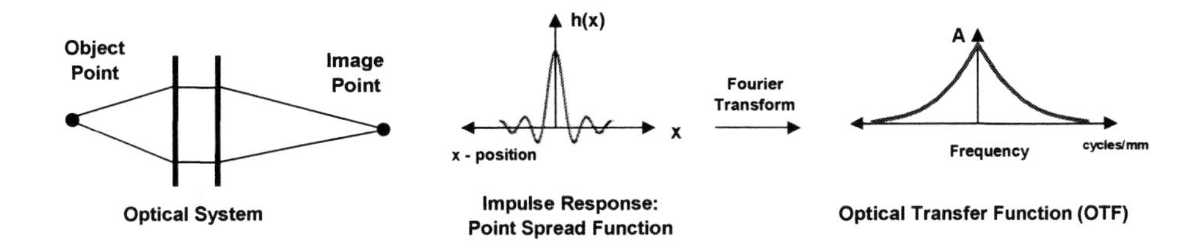

Figure 2.19 Optical system impulse response and transfer function.

2.5.1 Spatial domain

Computing the response of a physical system using the impulse response requires use of the convolution operation or convolution integral (also known as the Duhamel or superposition integral). This is a common method to compute the response of a mechanical system to an arbitrary time history. This same mathematical operation may be used to compute the image of any object by convolving the object with the PSF. An illustration of incoherent image formation for a periodic rectangle function is shown in Fig. 2.20. Note how the boundaries of the image are blurred as compared to the sharp boundaries of

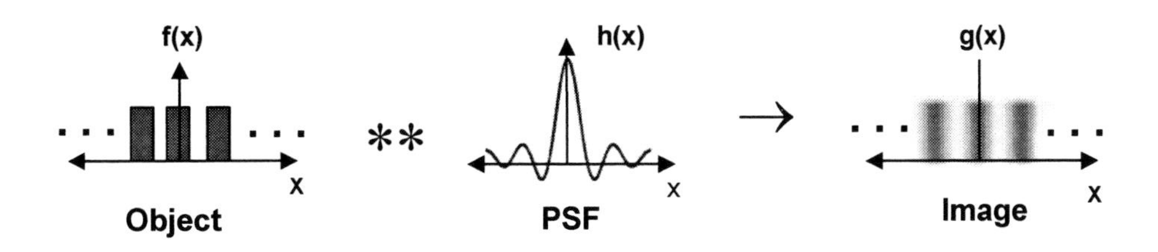

Figure 2.20 Image formation in the spatial domain (** represents the convolution operation).

the object. This is a consequence of the smoothing effect of the convolution operation. For an optical system to generate an image that is an exact duplicate of the object, the PSF would have to be a perfect point. As the size of the PSF increases, the smoothing effect increases and the quality of the image decreases. This should help explain why it is so important for high-performance optical systems to minimize the size of the blur diameter.

2.5.2 Frequency domain

A simple way to think of computing the response of a linear system in the frequency domain is to consider the physical system as acting like a frequency filter. For mechanical systems, loads that are a function of time are converted into harmonic frequency components in the temporal domain expressed in cycles per second, or Hz. For optical systems, objects are described by harmonic spatial frequency components typically expressed in cycles per millimeter.

The filtering aspect is dictated by the transfer function of the physical system. The filtering process is complex in that there are real and imaginary components. Think of the real part of the transfer function as an amplitude filter, and the imaginary part of the transfer function as a phase filter. The job of the transfer function is to determine the magnitude and relative phase of the response to each harmonic input. This is achieved by multiplying the transfer function by the harmonic input, which yields the spectral content of the output. An inverse Fourier transform is performed to convert back into the temporal or spatial domain.

For example, the image of a bar target of infinite extent, shown in Fig. 2.21, is computed using the frequency domain. The object is described using harmonic spatial frequencies computed using a Fourier series. The Fourier series representation of our object is given as

$$f(x) = \frac{A}{2} + \frac{2A}{\pi}\left[\cos 2\pi\xi x - \frac{1}{3}\cos 2\pi(3\xi)x + \frac{1}{5}\cos 2\pi(5\xi)x - \frac{1}{7}\cos 2\pi(7\xi)x + \ldots\right]. \tag{2.7}$$

Several of the individual frequency components are plotted and graphically summed to illustrate how spatial frequencies may be used to represent the object in Fig. 2.22. An abbreviated object spectrum is plotted in Fig. 2.23. The response or image of the optical system is controlled by the OTF, which tells us how the system will respond to each of the harmonic spatial frequencies that make up the object. The real part of the OTF or the amplitude filter tells us the magnitude of the response for each component. The amplitude filter is just the MTF, which was discussed in Sec. 2.4.4. The phase filter or *phase transfer function* (PTF) dictates the relative phase of each of the components.

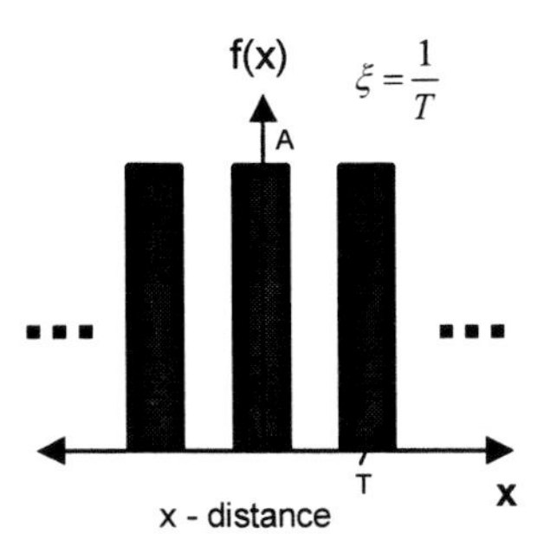

Figure 2.21 Bar target of infinite extent.

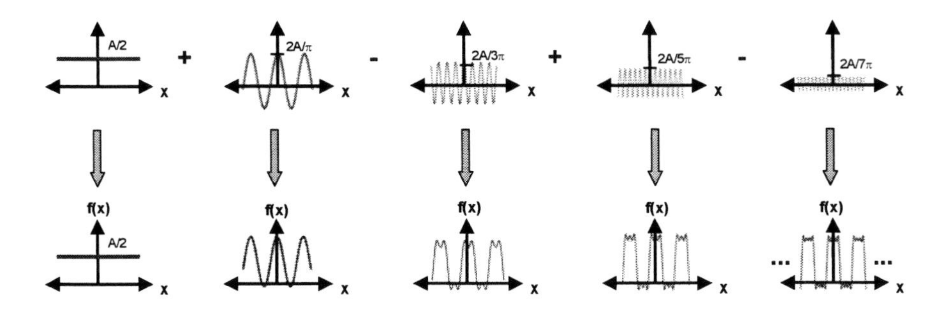

Figure 2.22 Harmonic spatial frequency components used to describe a bar target of infinite extent.

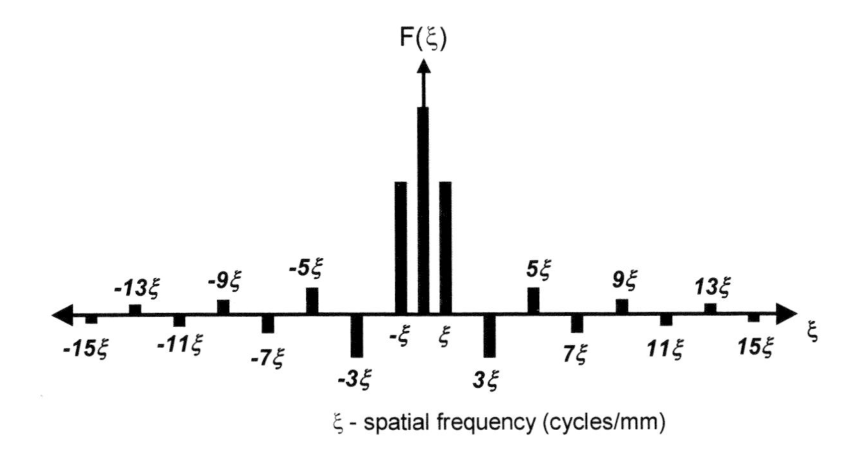

Figure 2.23 Spatial frequency content of a bar target of infinite extent.

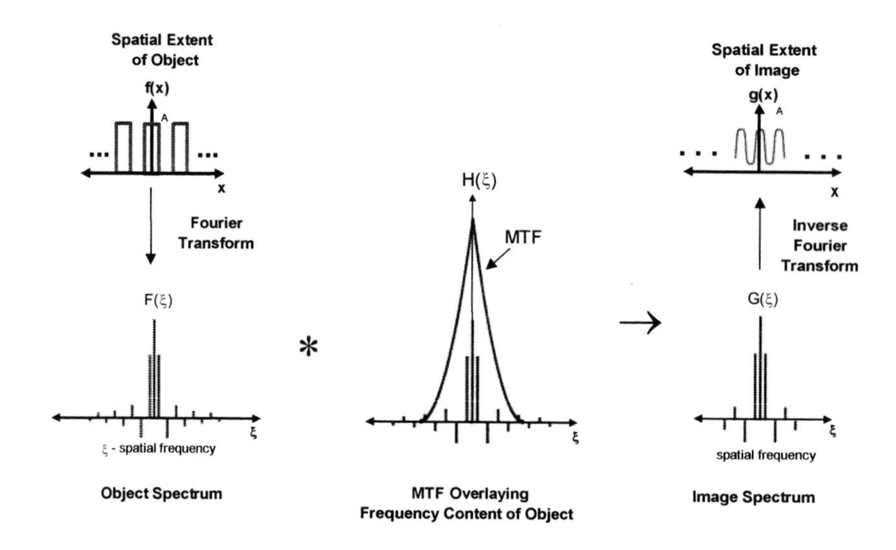

Figure 2.24 Image formation in the frequency domain.

Computationally, image formation is computed by multiplying the optical transfer function by the object spectrum as shown in Fig. 2.24. Note how the image spectrum is a truncated version of the object spectrum—a consequence of the filtering effect of the optical system. The higher-frequency components, which are responsible for the fine detail in the object, are cutoff. This leads to an image that is a rounded or smoothed version of the object. The image in the spatial domain (the domain where we can "see" the image) is computed by performing an inverse Fourier transform.

2.6 Zernike Polynomials

Zernike polynomials are useful in the interpretation of many types of optical surface data, including finite element derived surface deformations and wavefront maps due to thermal variations and mechanical stress. The Zernike polynomial represents a set of discrete data by a series of base surfaces each multiplied by a coefficient and summed. For example, surface data, Δ, may be represented by the polynomial series ϕ_i given in Eq. (2.8). The polynomials terms are summed with the contribution of each polynomial determined by the coefficients, a_i:

$$\Delta = \sum_i a_i \varphi_i = a_0 + a_1\varphi_1 + a_2\varphi_2 + a_3\varphi_3 + \ldots + a_i\varphi_i \,. \qquad (2.8)$$

The Zernike polynomial has several features that are particularly useful to optical systems. First, the polynomial set is orthogonal over a normalized circular aperture. Deleting terms from or adding terms to the set of polynomials does not affect the value of the remaining or the original coefficients. Second, the polynomials are minimized with respect to the rms departure from the reference surface using a least-squares fit. Thus, using a truncated set of Zernike terms to describe the data set or deleting terms from the Zernike set will always result in a decrease in the rms of the surface data. Finally, the Zernike terms may be related to one of the classical Seidel aberrations used to describe optical aberrations.

The Zernike polynomials are a complete set of polynomials with variables in radial, r, and azimuthal, θ, extent. A complete mathematical description for a given surface, $\Delta Z(r,\theta)$, is provided by Eq. (2.9), where A_{nm} and B_{nm} are the Zernike coefficients:

$$\Delta Z(r,\theta) = A_{00} + \sum_{n=2}^{\infty} A_{n0} R_n^0(r) + \sum_{n=1}^{\infty}\sum_{m=1}^{n} R_n^m \left[A_{nm}\cos(m\theta) + B_{nm}\sin(m\theta) \right]. \tag{2.9}$$

The radial dependence of the Zernike polynomials is given by the following expression:

$$R_n^m(r) = \sum_{s=0}^{\frac{n-m}{2}} (-1)^s \frac{(n-s)!}{s!\left(\frac{n+m}{2}-s\right)!\left(\frac{n-m}{2}-s\right)!} r^{(n-2s)}. \tag{2.10}$$

The variables n and m in Eqs. (2.9) and (2.10) are integer values and are known as the radial and circumferential wave number, respectively. There exist a few caveats in deriving the individual Zernike terms from the Zernike equations listed

m \ n	n = 0	1	2	3	4	5	6	7	8
0	Piston	0	Power	0	1-Sphr	0	2-Sphr	0	3-Sphr
1	0	Tilt	0	1-Coma	0	2-Coma	0	3-Coma	0
2	0	0	1-Astg	0	2-Astg	0	3-Astg	0	4-Astg
3	0	0	0	1-Tref	0	2-Tref	0	3-Tref	0
4	0	0	0	0	1-Tetr	0	2-Tetr	0	3-Tetr
5	0	0	0	0	0	1-Pent	0	2-Pent	0
6	0	0	0	0	0	0	1-Hexa	0	2-Hexa

1 = Primary	Sphr = Spherical	Tetr = Tetrafoil
2 = Secondary	Astg = Astigmatism	Pent = Pentafoil
3 = Tertiary	Tref = Trefoil	Hexa = Hexafoil
4 = Quaternary		

Figure 2.25 Zernike polynomials versus wave number.

above: $n - m$ must be an even number, and $n \geq m$. The Zernike polynomial as a function of the wave number is shown in Fig. 2.25.

There are two popular sets of Zernike polynomials that differ only in the way they are ordered. The Standard set[1] has an infinite number of terms; the first 37 terms are listed in Table 2.1. The Fringe set[2] is a reordered subset of the Standard Zernike terms, with a total of 37 terms, as listed in Table 2.2. The Fringe set includes higher-order radially symmetric terms while excluding the higher-order azimuthal terms.

Table 2.1 Standard Zernike polynomials (first 37 terms listed below).

	n	m	POLYNOMIAL	NAME
<1>	0	0	1	Piston
<2>	1	–1	$r\cos(\theta)$	X-Tilt
<3>	1	1	$r\sin(\theta)$	Y-Tilt
<4>	2	–2	$r^2\cos(2\theta)$	Pri Astigmatism-X
<5>	2	0	$2r^2 - 1$	Focus
<6>	2	2	$r^2\sin(2\theta)$	Pri Astigmatism-Y
<7>	3	–3	$r^3\cos(3\theta)$	Pri Trefoil-X
<8>	3	–1	$(3r^3 - 2r)\cos(\theta)$	Pri Coma-X
<9>	3	1	$(3r^3 - 2r)\sin(\theta)$	Pri Coma-Y
<10>	3	3	$r^3\sin(3\theta)$	Pri Trefoil-Y
<11>	4	–4	$r^4\cos(4\theta)$	Pri Tetrafoil-X
<12>	4	–2	$(4r^4 - 3r^2)\cos(2\theta)$	Sec Astigmatism-X
<13>	4	0	$6r^4 - 6r^2 + 1$	Pri Spherical
<14>	4	2	$(4r^4 - 3r^2)\sin(2\theta)$	Sec Astigmatism-Y
<15>	4	4	$r^4\sin(4\theta)$	Pri Tetrafoil-Y
<16>	5	5	$r^5\cos(5\theta)$	Pri Pentafoil-X
<17>	5	–3	$(5r^5 - 4r^3)\cos(3\theta)$	Sec Trefoil-X
<18>	5	–1	$(10r^5 - 12r^3 + 3r)\cos(\theta)$	Sec Coma-X
<19>	5	1	$(10r^5 - 12r^3 + 3r)\sin(\theta)$	Sec Coma-Y
<20>	5	3	$(5r^5 - 4r^3)\sin(3\theta)$	Sec Trefoil-Y
<21>	5	5	$r^5\sin(5\theta)$	Pri Pentafoil-Y
<22>	6	–6	$r^6\cos(6\theta)$	Pri Hexafoil-X
<23>	6	–4	$(6r^6 - 5r^4)\cos(4\theta)$	Sec Tetrafoil-X
<24>	6	–2	$(15r^6 - 20r^4 + 6r^2)\cos(2\theta)$	Ter Astigmatism-X
<25>	6	0	$20r^6 - 30r^4 + 12r^2 - 1$	Sec Spherical
<26>	6	2	$(15r^6 - 20r^4 + 6r^2)\sin(2\theta)$	Ter Astigmatism-Y
<27>	6	4	$(6r^6 - 5r^4)\sin(4\theta)$	Sec Tetrafoil-Y
<28>	6	6	$r^6\sin(6\theta)$	Pri Hexafoil-Y
<29>	7	–7	$r^7\cos(7\theta)$	Pri Septafoil-X
<30>	7	–5	$(7r^7 - 6r^5)\cos(5\theta)$	Sec Pentafoil-X
<31>	7	–3	$(21r^7 - 30r^5 + 10r^3)\cos(3\theta)$	Ter Trefoil-X
<32>	7	–1	$(35r^7 - 60r^5 + 30r^3 - 4r)\cos(\theta)$	Ter Coma-X
<33>	7	1	$(35r^7 - 60r^5 + 30r^3 - 4r)\sin(\theta)$	Ter Coma-Y
<34>	7	3	$(21r^7 - 30r^5 + 10r^3)\sin(3\theta)$	Ter Trefoil-Y
<35>	7	5	$(7r^7 - 6r^5)\sin(5\theta)$	Sec Pentafoil-Y
<36>	7	7	$r^7\sin(7\theta)$	Pri Septafoil-Y
<37>	8	8	$r^8\cos(8\theta)$	Pri Octafoil-X

Table 2.2 Fringe Zernike polynomials.

	n	m	POLYNOMIAL	NAME
<1>	0	0	1	Piston
<2>	1	–1	$r\cos(\theta)$	Tilt-X
<3>	1	1	$r\sin(\theta)$	Tilt-Y
<4>	2	0	$2r^2 - 1$	Focus
<5>	2	–2	$r^2\cos(2\theta)$	Pri Astig.-X
<6>	2	2	$r^2\sin(2\theta)$	Pri Astig.-Y
<7>	3	–1	$(3r^3 - 2r)\cos(\theta)$	Pri Coma-X
<8>	3	1	$(3r^3 - 2r)\sin(\theta)$	Pri Coma-Y
<9>	4	0	$6r^4 - 6r^2 + 1$	Pri Spherical
<10>	3	–3	$r^3\cos(3\theta)$	Pri Trefoil-X
<11>	3	3	$r^3\sin(3\theta)$	Pri Trefoil-Y
<12>	4	–2	$(4r^4 - 3r^2)\cos(2\theta)$	Sec Astig.-X
<13>	4	2	$(4r^4 - 3r^2)\sin(2\theta)$	Sec Astig.-Y
<14>	5	–1	$(10r^5 - 12r^3 + 3r)\cos(\theta)$	Sec Coma-X
<15>	5	1	$(10r^5 - 12r^3 + 3r)\sin(\theta)$	Sec Coma-Y
<16>	6	0	$20r^6 - 30r^4 + 12r^2 - 1$	Sec Spherical
<17>	4	4	$r^4\cos(4\theta)$	Pri Tetrafoil-X
<18>	4	4	$r^4\sin(4\theta)$	Pri Tetrafoil-Y
<19>	5	–3	$(5r^5 - 4r^3)\cos(3\theta)$	Sec Trefoil-X
<20>	5	3	$(5r^5 - 4r^3)\sin(3\theta)$	Sec Trefoil-Y
<21>	6	–2	$(15r^6 - 20r^4 + 6r^2)\cos(2\theta)$	Ter Astig.-X
<22>	6	2	$(15r^6 - 20r^4 + 6r^2)\sin(2\theta)$	Ter Astig.-Y
<23>	7	–1	$(35r^7 - 60r^5 + 30r^3 - 4r)\cos(\theta)$	Ter Coma-X
<24>	7	–1	$(35r^7 - 60r^5 + 30r^3 - 4r)\sin(\theta)$	Ter Coma-Y
<25>	8	0	$70r^8 - 140r^6 + 90r^4 - 20r^2 + 1$	Ter Spherical
<26>	5	5	$r^5\cos(5\theta)$	Pri Pentafoil-X
<27>	5	5	$r^5\sin(5\theta)$	Pri Pentafoil-Y
<28>	6	4	$(6r^6 - 5r^4)\cos(4\theta)$	Sec Tetrafoil-X
<29>	6	4	$(6r^6 - 5r^4)\sin(4\theta)$	Sec Tetrafoil-Y
<30>	7	–3	$(21r^7 - 30r^5 + 10r^3)\cos(3\theta)$	Ter Trefoil-X
<31>	7	3	$(21r^7 - 30r^5 + 10r^3)\sin(3\theta)$	Ter Trefoil-Y
<32>	8	–2	$(56r^8 - 105r^6 + 60r^4 - 10r^2)\cos(2\theta)$	Qua Astig.-X
<33>	8	2	$(56r^8 - 105r^6 + 60r^4 - 10r^2)\sin(2\theta)$	Qua Astig.-Y
<34>	9	–1	$(126r^9 - 280r^7 + 210r^5 - 60r^3 + 5r)\cos(\theta)$	Qua Coma-X
<35>	9	1	$(126r^9 - 280r^7 + 210r^5 - 60r^3 + 5r)\sin(\theta)$	Qua Coma-Y
<36>	10	0	$252r^{10} - 630r^8 + 560r^6 - 210r^4 + 30r^2 - 1$	Qua Spherical
<37>	12	0	$924r^{12} - 2772r^{10} + 3150r^8 - 1680r^6 + 420r^4 - 42r^2 + 1$	Qin Spherical

The first term of the Zernike series, piston, is just a dc-term representing a bias to the original data. The tilt terms represent perpendicular planes. Focus represents a quadratic or parabolic change in the radial extent of the surface shape. Astigmatism is best described as the shape of a horse's saddle, or a potato chip, possessing unequal curvatures along perpendicular axes. Coma is a surface with a pair of humps, where one of the humps is inverted. Three-dimensional contour plots of several Zernike polynomials are shown in Fig. 2.26.

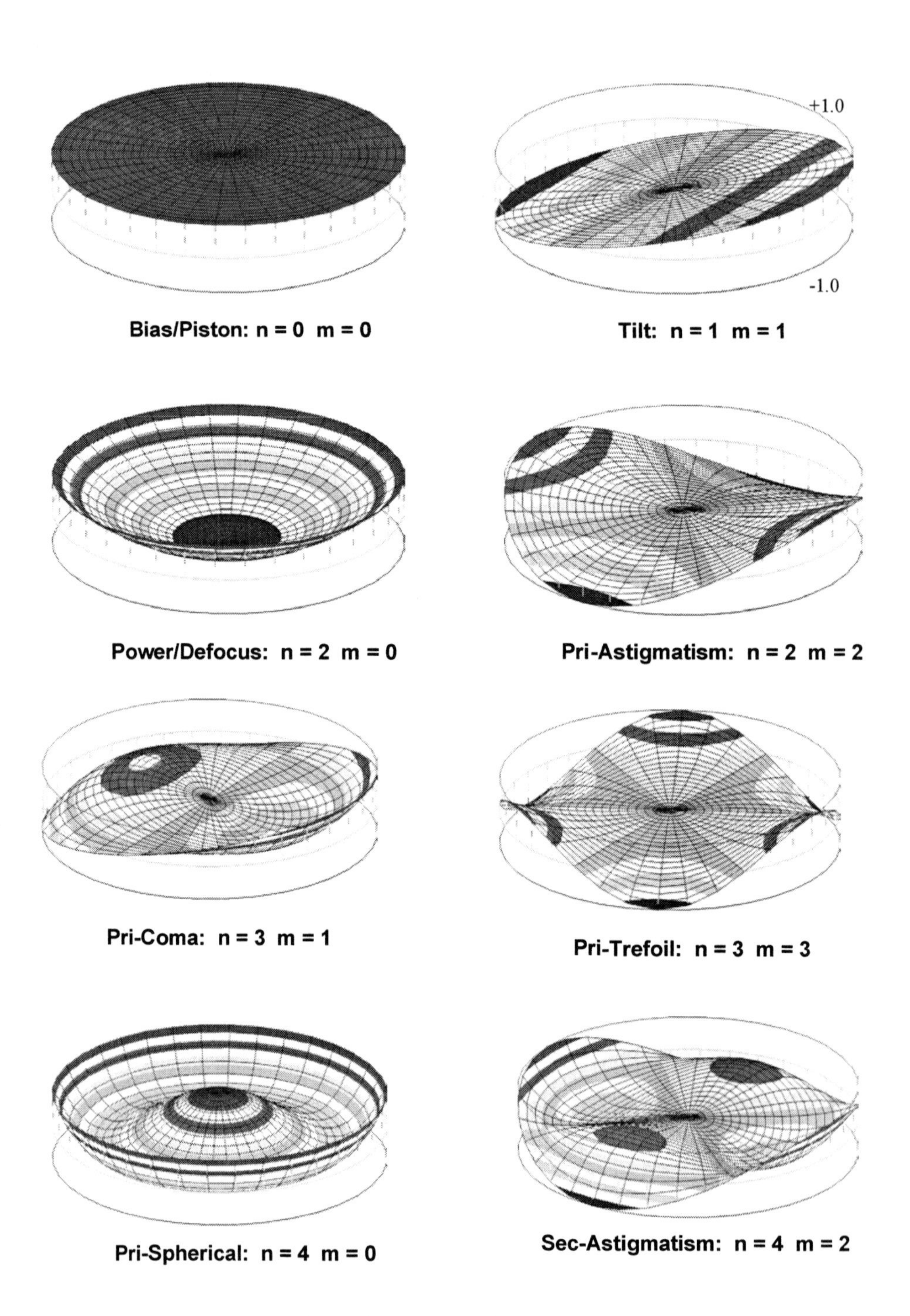

Figure 2.26 Zernike polynomials (continued, next page).

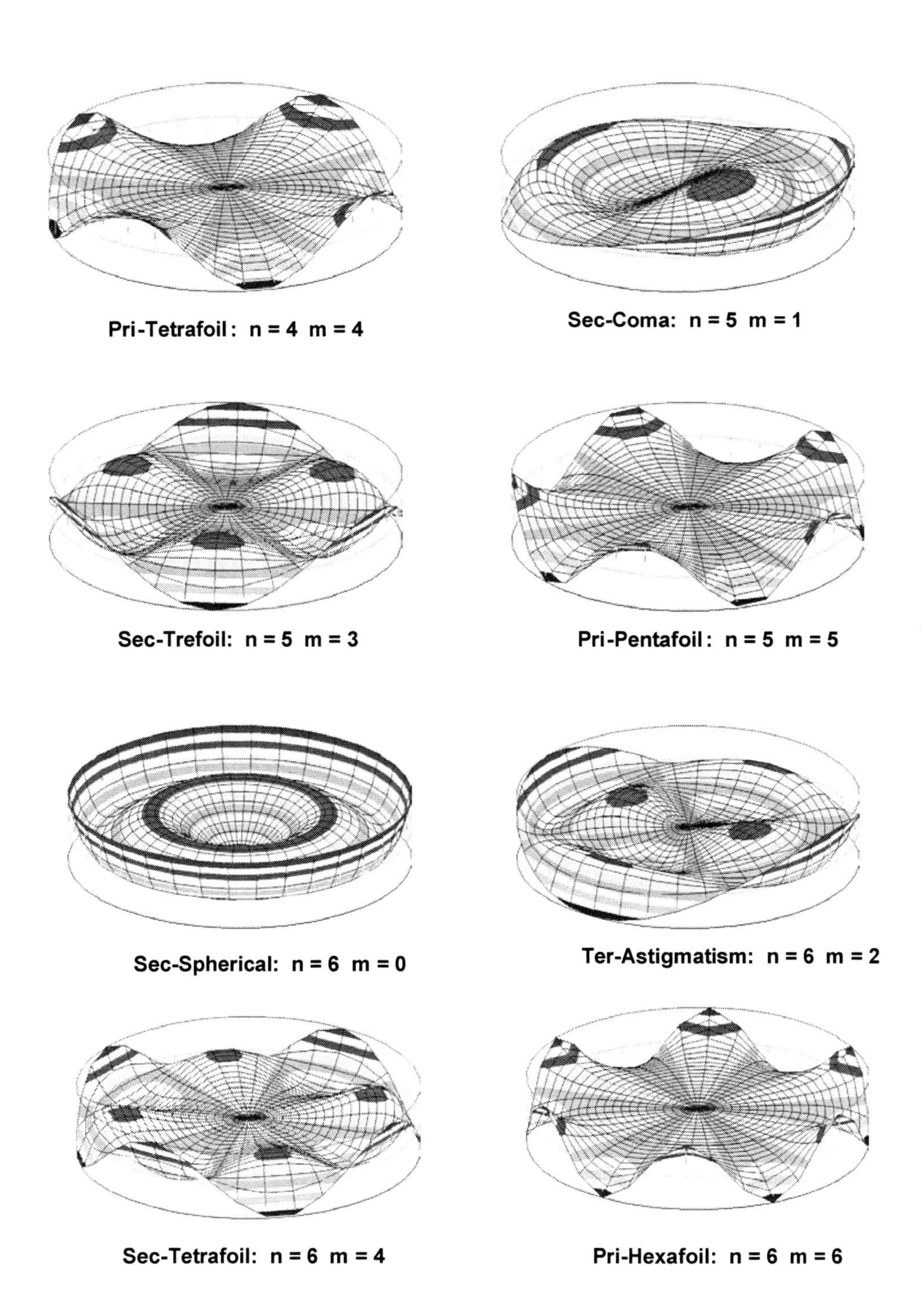

Figure 2.26 Continued.

2.6.1 Magnitude and phase

Each of the Zernike terms that are a function of the azimuthal angle θ, such as tilt and astigmatism, have a cosine and sinusoidal dependence, represented by the Zernike coefficients A_{nm} and B_{nm}. Each pair of these terms may be expressed as a single term with an associated magnitude and phase as given below:

$$M = \sqrt{A_{nm}^2 + B_{nm}^2}\ , \tag{2.11}$$

$$\text{Phase} = \frac{1}{m}\tan^{-1}\frac{B_{nm}}{A_{nm}}. \tag{2.12}$$

The phase may be thought as the azimuthal orientation. This alternative format, listed in Table 2.3, reduces the Fringe Zernike terms from 37 to 22 helping to simplify data interpretation.

Table 2.3 Combined Fringe Zernike terms may be represented as magnitude and phase.

ORDER TERM	N	M	ABERRATION TYPE
1	0	0	Piston
2	1	1	Tilt
3	2	0	Focus
4	2	2	Pri Astigmatism
5	3	1	Pri Coma
6	4	0	Pri Spherical
7	3	3	Pri Trefoil
8	4	2	Sec Astigmatism
9	5	1	Sec Coma
10	6	0	Sec Spherical
11	4	4	Pri Tetrafoil
12	5	3	Sec Trefoil
13	6	2	Ter Astigmatism
14	7	1	Ter Coma
15	8	0	Ter Spherical
16	5	5	Pri Pentafoil
17	6	4	Sec Tetrafoil
18	7	3	Ter Trefoil
19	8	2	Qua Astigmatism
20	9	1	Qua Coma
21	10	0	Qua Spherical
22	12	0	Qin Spherical

2.6.2 Orthogonality of Zernike polynomials

The condition of orthogonality allows individual Zernike terms to be subtracted or added to the polynomial series without changing the value of the other coefficients. Orthogonality is met only for continuous data. It is approximate for data uniformly spaced, and degrades significantly as the data becomes irregular. One method to check orthogonality is to fit the data to a varying number of Zernike terms. If the values of the coefficients change significantly then the Zernike terms are not orthogonal. This section discusses the orthogonality of Zernike polynomials in representing continuous data and for describing uniformly and irregularly spaced data, typical of optical surface finite element meshes.[3]

Mathematically, two functions, Φ_1 and Φ_2, representing continuous data, are orthogonal over a unit circle if:

$$\int_0^1 \int_0^{2\Pi} \Phi_1 \Phi_2 \rho d\Theta d\rho = 0, \tag{2.13}$$

which simply means that the area of the product of two functions over the unit circle is zero. For axisymmetric functions, the above equation reduces to

$$2\Pi \int_0^1 \Phi_1 \Phi_2 \rho d\rho = 0. \tag{2.14}$$

It may be shown using the above equations, that each of the Zernike functions is orthogonal to each other over a unit circle. For example, the orthogonality of the piston and focus terms and the focus and spherical terms is verified in Eqs. (2.15) and (2.16), respectively.

$$2\Pi \int_0^1 (1)\left(2\rho^2 - 1\right)\rho d\rho = 2\Pi\left(\frac{2}{4} - \frac{1}{2}\right) = 0, \tag{2.15}$$

$$2\Pi \int_0^1 \left(2\rho^2 - 1\right)\left(6\rho^4 - 6\rho^2 + 1\right)\rho d\rho = 2\Pi\left(\frac{12}{8} - \frac{18}{6} + \frac{8}{4} - \frac{1}{2}\right) = 0. \tag{2.16}$$

In surveying the various Zernike polynomial expressions, it may be observed that each of the higher-order polynomials contains an appropriate amount of the lower-order polynomial. For example, spherical aberration contains both focus and piston terms. The addition of the lower-order terms allows orthogonality to be maintained.

2.6.2.1 Noncircular apertures

Fitting Zernike polynomials to data over noncircular apertures requires that the pupil be sized to the radius that encloses the full area of the aperture. This is shown for an elliptical and obscured aperture in Fig. 2.27. Within the full pupil radius there will be points in which no data exists, resulting in loss of orthogonality. (The accuracy of the polynomial representation, however, is no less accurate).

For example, consider the primary mirror of a Cassegrain telescope that includes a central hole of $\rho = 0.2$. The orthogonality of the Zernike terms is now lost as demonstrated using the piston and focus terms below:

$$2\Pi \int_{0.2}^{1} (1)\left(2\rho^2 - 1\right)\rho d\rho = 2\Pi\left[\left(\frac{2}{4}-\frac{1}{2}\right)-\left(\frac{.0032}{4}-\frac{.04}{2}\right)\right] = 0.12 \neq 0. \qquad (2.17)$$

Orthogonality is also lost on a noncircular geometry, such as a square optic:

$$\int_{-1}^{+1}\int_{-1}^{+1} (1)\left(2\rho^2 - 1\right) dxdy \neq 0. \qquad (2.18)$$

Variations of the Zernike polynomials do exist that are orthogonal over noncircular apertures.[4] However, their discussion is beyond the scope of this text.

Figure 2.27 Noncircular apertures: elliptical and obscured.

2.6.2.2 Discrete data

The orthogonality of Zernike polynomials is also lost when fitting terms to discrete data. For discrete data evaluated at node k, equation (2.13) becomes

$$\sum_k \Phi_{1k}\Phi_{2k}A_k = 0, \tag{2.19}$$

where A_k is the area associated with node k. A comparison of orthogonality using numerical integration for the piston, focus, and spherical terms fit to varying mesh densities is shown in Table 2.4 where the residual error verses the number of equally spaced radial integration points, K, over the unit circle are listed. The diagonal terms $\Phi_j\Phi_j$ represent the square of the rms, and the off-diagonal terms $\Phi_i\Phi_j$ represent the coupling or nonorthogonality. Notice as the number of radial node points increase in the mesh density, the polynomial terms become increasingly orthogonal.

Table 2.4: Numerical integration on a unit circle.

K	$\Phi_0\Phi_0$	$\Phi_1\Phi_1$	$\Phi_2\Phi_2$	$\Phi_0\Phi_1$	$\Phi_0\Phi_2$	$\Phi_1\Phi_2$
10	1.0000	.34660	.23838	.00500	.01990	.02460
20	1.0000	.33666	.20990	.00125	.00499	.00623
50	1.0000	.33387	.20160	.00020	.00080	.00100
100	1.0000	.33347	.20040	.00005	.00020	.00025
200	1.0000	.33337	.20010	.00001	.00005	.00006
500	1.0000	.33334	.20002	.00000	.00001	.00001
1000	1.0000	.33333	.20000	.00000	.00000	.00000

Orthogonality is also a function of the uniformity of the data. For instance, coupling of the Zernike terms increases when the finite element mesh is nonuniform. A comparison of a uniform isomesh and irregular "automesh" is shown in Fig. 2.28.

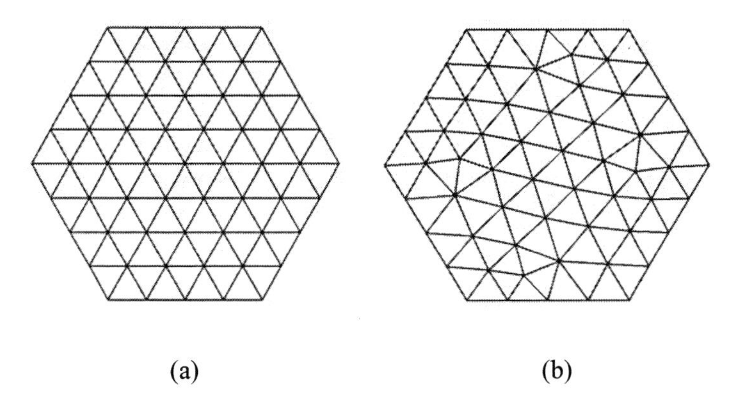

(a) (b)

Figure 2.28 (a) Regular "isomesh" model, and (b) irregular "automesh" model.

In the isomesh, axisymmetric terms are coupled only to other axisymmetric terms. In the irregular mesh, axisymmetric terms pick up additional coupling with the nonaxisymmetric terms such as astigmatism, coma, and trefoil.

2.6.3 Computing the Zernike polynomial coefficients

Fitting Zernike polynomials to a set of surface data may be performed using a least-squares fit. This section assumes that the surface data is computed using finite element analysis, although any two-dimensional data may be fit using this technique.[5] Consider a grid of node points, i, representing an optical surface in a finite element model. A least-squares error function is defined as the difference between the polynomial description of the deformation, Z_i, and the actual finite element computed deformation, δ_i. A weighting function, W_i, may be applied that is proportional to the area that each node point represents on the optical surface. This accounts for the variation in nodal density and allows for an equitable contribution of each node point to the overall computation. The area may be computed as a fraction of the total normal surface area or projected surface area. Typically, use of the projected area yields a more representative fit as use of the normal surface area increases the contribution of the nodes on the edges of the surface. The least-squares error function, E, is given as

$$E = \sum W_i (\delta_i - Z_i)^2. \tag{2.20}$$

The Zernike polynomial approximation, Z_i, is given by the summation of the Zernike coefficients, c_j, which are being solved, and the Zernike polynomial, ϕ_{ij}:

$$Z_i = \sum c_j \phi_{ji}. \tag{2.21}$$

This yields the following least-squares error function:

$$E = \sum W_i \left(\delta_i - \sum c_j \phi_{ji}\right)^2. \tag{2.22}$$

To compute the best-fit Zernike coefficients, the error function is minimized with respect to the coefficients. This is done mathematically by taking the derivative of the error function with respect to the coefficients and setting it equal to zero:

$$\frac{\partial E}{\partial c_j} = 2\sum W_i \left(\delta_i - \sum c_j \phi_{ji}\right) \phi_{ji} = 0. \tag{2.23}$$

The resulting expression is in linear matrix form allowing the coefficients, $\{c\}$, to be solved using Gauss elimination:

$$[H]\{c\} = \{p\}, \tag{2.24}$$

where

$$p_j = \sum W_i \delta_i \phi_{ji}, \tag{2.25}$$

and

$$H_{jk} = \sum W_i \phi_{ji} \phi_{ki}. \tag{2.26}$$

Once the Zernike coefficients have been computed, the rms fit error should be computed to determine how well the polynomial represents the actual data. The rms fit error is computed as the rms of the difference between the polynomial representation and the actual data. The required accuracy depends on the specific application but, generally, the rms fit error should be a small fraction of the rms surface error.

2.7 Legendre-Fourier Polynomials

Legendre-Fourier polynomials form an orthogonal set of surface descriptors for cylindrical optics.[6,7] A notable example of an optical system using cylindrical optics is NASA's Chandra X-Ray Observatory. The Legendre-Fourier polynomials are a product of two sets of functions where the Legendre polynomial represents the axial direction, and the Fourier series represents the azimuthal direction. The mathematical description of the Legendre-Fourier polynomials, $f(z,\theta)$, is shown below, where a_{nm} are the coefficients and G_{nm} are the polynomials:

$$f(z,\theta) = \sum_{n=0}^{\infty} \left[a_{n0} G_{n0} + \sum_{m=1}^{\infty} \left(a_{nm}^C G_{nm}^C + a_{nm}^S G_{nm}^S \right) \right], \tag{2.27}$$

where

$$G_{n0}(z,\theta) = \sqrt{2n+1} P_n(z), \tag{2.28}$$

$$G_{nm}^C(z,\theta) = \sqrt{2(2n+1)} P_n(z) \cos(m\theta), \tag{2.29}$$

and

$$G_{nm}^S(z,\theta) = \sqrt{2(2n+1)} P_n(z) \sin(m\theta). \tag{2.30}$$

Several of the Legendre-Fourier polynomials base surfaces are shown in Fig. 2.29. Azimuthally symmetric terms along with decenter, tilt, and out-of-roundness are illustrated.

Mathematically, the Legendre polynomials, $P_n(z)$, are expressed as

$$P_n(x) = \sum_{k=0}^{K} (-1)^k \frac{(2n-2k)!}{2^n k!(n-k)!(n-2)!} x^{n-2k} . \tag{2.31}$$

A truncated set is plotted in Fig. 2.30. Fitting the Legendre-Fourier polynomials to cylindrical surface data is accomplished in the same manner as fitting the Zernike polynomials.

2.8. Aspheric and x-y polynomials

Aspheric and x-y polynomials are additional polynomial sets often used to describe optical surface data as expressed in Eqs. (2.32) and (2.33), respectively:

$$z = A_0 + A_1 r + A_2 r^2 + \ldots + A_n r^n \tag{2.32}$$

$$z = A_{00} + A_{10} x + A_{01} y + A_{20} x^2 + A_{11} xy + A_{02} y^2 + \ldots + A_{nm} x^n y^m \tag{2.33}$$

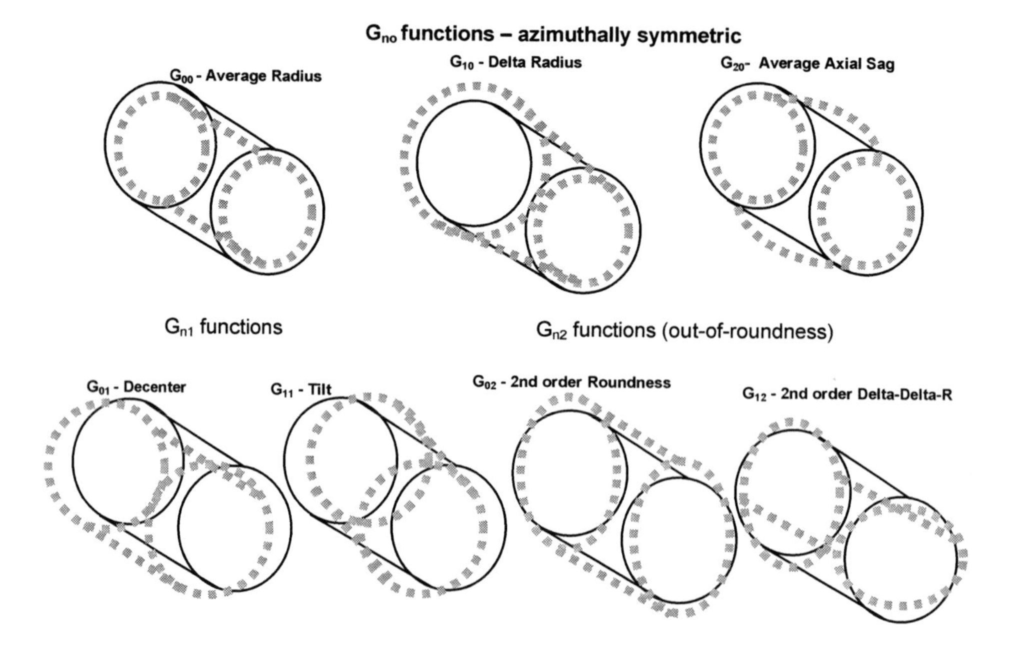

Figure 2.29 Legendre-Fourier polynomials.

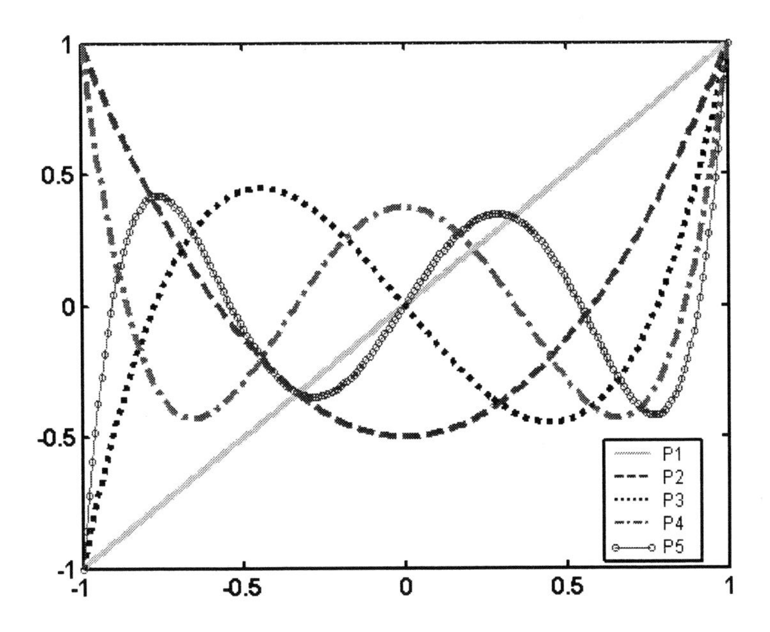

Figure 2.30 Legendre polynomials.

For example, x-y polynomials are useful to represent the surface deformations of a mirror with rectangular stiffening ribs. Aspheric polynomials may be used to fit axisymmetric behavior to a very high order.

References

1. Born, M. and E. Wolf, *Principles of Optics*, Pergamon Press, New York, (1964).
2. Wyatt, J. C., Creath, K., "Basic wavefront aberration theory for optical metrology," *Applied Optics and Optical Engineering*, Vol. XI.
3. Genberg, V. L., Michels, G. J., Doyle, K. B., "Orthogonality of Zernike Polynomials," *Proceedings of SPIE*, **4771**, Bellingham, WA (2002).
4. Swantner, W., Chow, W.W., "Gram-Schmidt orthonormalization of Zernike polynomials for generalized aperture shapes," Applied Optics, 33(10) (1994).
5. Genberg, V. L., "Optical surface evaluation," Proceedings of SPIE, **450,** Bellingham, WA (1983).
6. Genberg, V. L., "Structural Analysis of Optics" *Handbook of Optomechanical Engineering*, CRC Press (1997).
7. Glenn, P., "Set of orthonormal surface error descriptors for near cylindrical optics," *J. Optical Engineering*, **23**(4) (1984).

⋞Chapter 3⋟ Optomechanical Displacement Analysis Methods

This chapter presents guidelines relevant to finite element model construction and analysis methods for predicting the motion and deformation of optics and optical structures. A key idea to be conveyed is that an analyst's choice of how to model optical components is dependent on several factors. The most obvious factor, of course, is that the mechanical behavior of the hardware will require that certain modeling features and methods be used in order to accurately predict a system's true behavior. However, consideration of this factor alone would lead us to construct finite element models that capture the stress states of every fillet and stress riser in the system. This approach is certainly not practical when schedule and cost constraints are prohibitive of such an effort. Fortunately, predicting most optomechanical performance metrics do not require models capable of such extensive mechanical representation. Often, only first-order mechanical behavior is needed to provide sufficient accuracy in the prediction of optical performance. Yet another important factor in the choice of a modeling method is how the analysis results will be used. For example, if the goal of an analysis is to compare several design concepts in the early phases of a feasibility study, then simple models, which may not accurately predict the absolute behavior, may nevertheless be effective in providing relative performance predictions among the various design concepts. By presenting an array of modeling methods each with their own limitations and strengths, it is hoped that the reader becomes better able to make the best modeling decisions to meet the technical, schedule, and cost requirements of any optomechanical displacement analysis task.

3.1 Displacement Models of Optics

3.1.1 Definitions

In discussing the displacement models of optics, it is helpful to define a few terms relevant to optic motion and deformation. These definitions aid future discussions of the limitations of the various modeling methods.

Component rigid-body motion is the set of average translations and rotations of the optical component. This quantity can also be thought of as the motion of the center of mass of the optical component as illustrated in Fig. 3.1.

Optical surface rigid-body motion is the set of average translations and rotations of the optical surface of an optical component. Figure 3.1 illustrates that this motion may be different from the component rigid-body motion.

Global surface deformation is the component of the total surface deformation that is exhibited over most or the entire optical surface. Such deformations are well predicted, in general, by both coarse and detailed finite element models and are reasonably approximated by low-order surface polynomials.

Local surface deformation is the component of total surface deformation of an optical surface that is confined to local regions. Such deformations generally require more detailed finite element models to be accurately predicted. Local surface deformations also require very high order surface polynomials to be described, or they may not be representable by polynomials at all. Such deformations usually result from mount-induced effects.

Quilting deformation is a specific type of local surface deformation seen in lightweighted mirrors that have a relatively thin optical facesheet backed by a cellular core structure. Sources of quilting deformation include thermoelastic deformation of the optical facesheet caused by nonuniform thermal gradients through the thickness of the optical facesheet and elastic deformation of the optical facesheet due to an applied gravity load or polishing pressure.

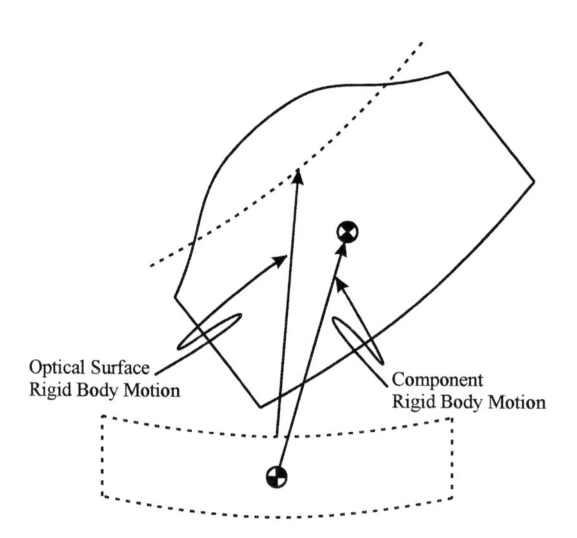

Figure 3.1 Component rigid-body motion and optical surface rigid-body motion are distinct quantities.

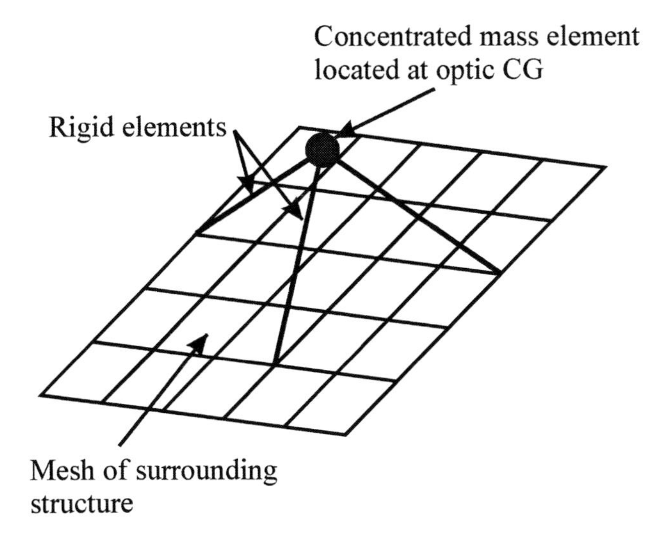

Figure 3.2 Single-point model of an optic connected to a surrounding mesh with rigid elements.

3.1.2 Single-point models

The simplest of all displacement models is the single-point model where the optic is represented by a single node as shown in Fig. 3.2. In such a model only the component rigid-body motions are predicted. Therefore, such a model is used when the elastic deformation of the optic is not important to the goal of the analysis. Common applications for single-point optic models are for small mirrors and lenses whose elastic deformations do not significantly contribute to optical performance degradation. It is also assumed that the optical surface rigid-body motion can be approximated by the component rigid-body motion, or it is not of interest to results of the analysis. If thermoelastic effects or other mechanical behaviors cause the component rigid-body motion to be measurably different from the optical surface rigid-body motion as shown in Fig. 3.1, a model of this type may not be acceptable.

The connection of the single-point model to its supporting structure is an important consideration. The selected element, or elements, used to perform the connection may be rigid or elastic. Rigid-element formulations, however, have no thermal expansion capabilities in most commercially available finite element codes. Therefore, erroneous states of stress can be predicted by a rigidly connected single-point model of an optic if thermoelastic loads are applied. To model such a stiff connection in the presence of thermoelastic loads it is recommended that beam elements with stiff properties and a proper thermoelastic expansion definition be used to model the connection. In addition, the use of zero-length rigid elements or pin flags at the connection points to the supporting structure may be used to link the single-point model in only specific degrees of freedom. Such a connection may be used to represent a kinematic interface.

When single-point models are used in dynamics analyses, it is important to include a complete description of the optic's mass. This mass description should include mass moments and mass products of inertia in addition to the translational mass. Such mass properties are defined on a concentrated mass element available in most finite element codes. These mass properties can be computed from analytical equations for simple geometries or by solid modeling tools for more complicated shapes. Analytical equations for simple solid geometries can be found in most mechanical design, vibrations, or dynamics textbooks.

Therefore, while single-point models are limited in the output they provide, they can be an excellent choice for including the mass of an optic and predicting its component rigid-body motion. In addition, single-point models are very easy to alter, making them excellent tools for early design trades and concept studies.

3.1.3 Solid optics

Solid optics are characterized by geometric topology that lacks lightweighting or discrete stiffening. Examples are lenses, solid mirrors, and windows.

3.1.3.1 Two-dimensional models of solid optics

Some solid optics exhibit mechanical behavior that can be well approximated under the assumptions of plate or shell behavior. In such cases, the elastic stiffness of a 2D solid optic model is defined by membrane, bending, and transverse-shear stiffnesses. The dimensional parameters on which these stiffnesses depend are the thickness of the optic and the transverse shear factor. The transverse shear factor of a homogeneous plate is 0.8333.

Two-dimensional models can provide excellent predictions of global elastic behavior for static and vibration analyses. An important limitation of 2D-element optic models, however, is that they do not predict deformation effects in the direction through the thickness of the optic. Therefore, their rigid-body motions and global elastic deformations are representative of the midplane of the optic and not necessarily that of the optical surface. Differences between the behavior of the midplane of an optic and its optical surface can be caused by mount-induced loads and thermoelastic growth through the thickness of the optic. Furthermore, mount-induced loads will show greater local deformations in 2D-element models than may actually exist at the optical surface of the actual hardware. Therefore, the analyst should choose this method of modeling a solid optic only when it is reasonable to assume that such effects are not significant to the overall goal of the analysis.

Plate-element meshes can also be used to model components for which displacement predictions are not required but are included to provide a reasonable representation of stiffness. Figure 3.3(a) shows a lens to be modeled as part of a lens barrel model. However, suppose displacements are not required of the lens shown in the figure. A model that correctly represents its stiffness

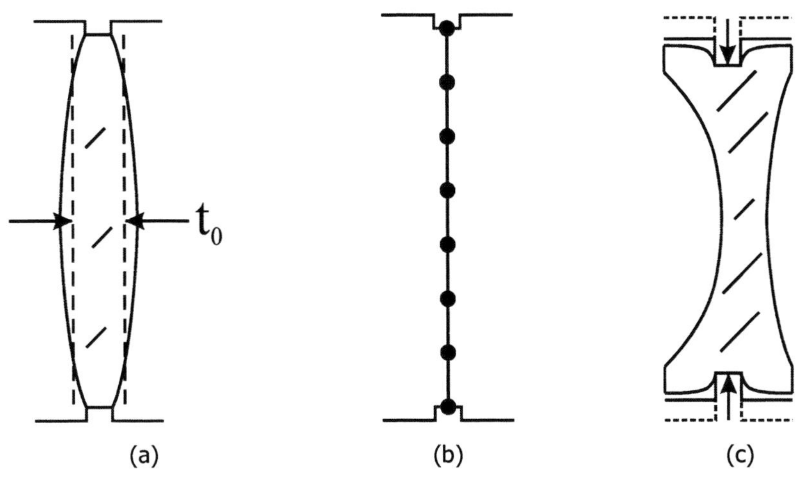

Figure 3.3 Modeling of lenses with 2D-models: (a) lens with relatively constant thickness, t_0, (b) corresponding 2D-element mesh, and (c) elastic behavior in a lens which would not be represented by a 2D element mesh.

may be required to obtain useable displacement results elsewhere in the system. The lens has a relatively constant thickness approximately equal to t_0 as shown in Fig. 3.3(a), and can be reasonably represented by the plate mesh shown in Fig. 3.3(b). The stiffness of the lens shown in Fig. 3.3(c), however, may not be well represented by a plate mesh due to the inability of such a model to predict potential deformations such as those shown. Such a model may have to be constructed of 3D solid elements, as described in the next section, in order to provide a reasonable approximation of its stiffness.

3.1.3.2 Three-dimensional element models of solid optics

Components whose elastic behavior cannot be accurately represented by plate assumptions require solid-element formulations that use the full 3D representation of Hooke's law. Examples of such components are thick lenses, thick solid mirrors, and prisms. Figure 3.4 shows some examples of such models.

The construction of solid-element models deserves a few guidelines to be followed in most cases. Solid-element models of lenses and mirrors should have at least four trilinear elements through their thicknesses. Such a minimum resolution is required in most cases to provide a reasonably accurate prediction of the variation in stress states through the thickness of the component. In many cases, more than four elements will be required. The number of elements required is dictated by the variation of displacements through the thickness of the component and the elements' ability to represent them.

The use of automeshing algorithms to generate meshes of highly symmetric optical components as shown in Fig. 3.4 is not desirable. Automeshing routines will commonly generate nonsymmetrical meshes for even the most symmetric

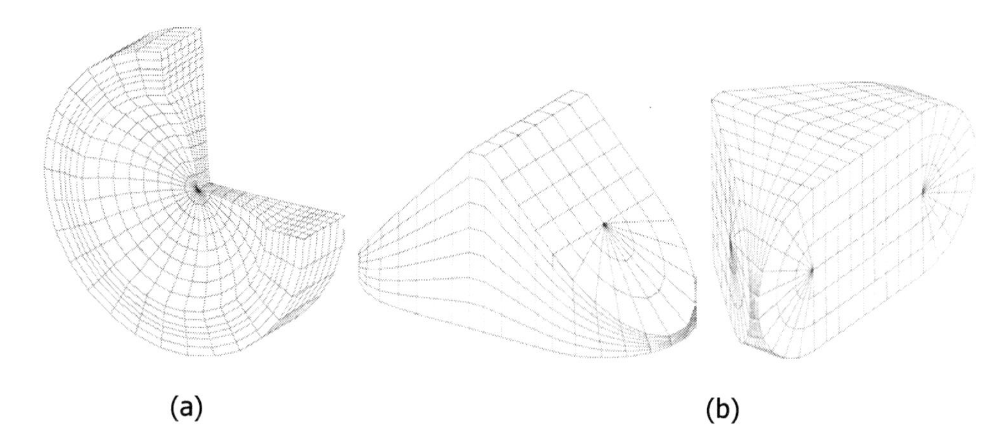

Figure 3.4 Examples of 3D solid models: (a) lens and (b) Porro prism.

structures. Such asymmetries in element meshes can generate nonsymmetrical results for problems with symmetric behavior. Automeshing routines, on the other hand, are not without usefulness—they can be useful in situations involving very complicated geometry not meshable by six-sided and five-sided solid elements.

The use of the four-noded constant-strain tetrahedron element should be strictly avoided. The formulation of this element assumes a constant state of strain throughout its volume, resulting in a mesh that is too stiff for useable displacement results. Often it is tempting for users to resort to this element due to its use in automeshing algorithms. However, if tetrahedron elements must be used, then ten-noded tetrahedron elements should be employed.

3.1.4 Lightweight mirror models

Modeling of lightweight mirrors is a specific application that deserves special attention. Three types of lightweight mirror displacement models are discussed in this section. Each type of model has its own strengths and weaknesses, and the analyst is encouraged to keep the goals of the analysis in mind while choosing which type of model to use.

A lightweight mirror may have one of the various core-cell shapes as shown in the silicon carbide mirrors in Fig. 3.5. In addition, lightweight mirrors may include a back facesheet that provides increased plate bending stiffness. Lightweight mirrors are also fabricated with varying diameter-to-depth ratios to suit particular applications. All of these constructions can be modeled by the techniques discussed in this section. Of course, nonoptical structures similar in construction to lightweight mirrors may also be modeled by these methods.

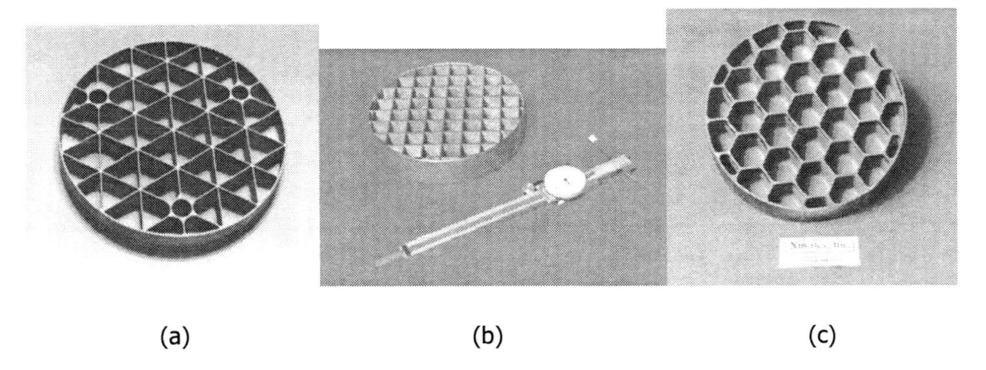

Figure 3.5 Examples of lightweight mirror construction: (a) triangular core, (b) square core, and (c) hexagonal core. (Courtesy of Xinetics, Inc., Devens, Massachusetts.)

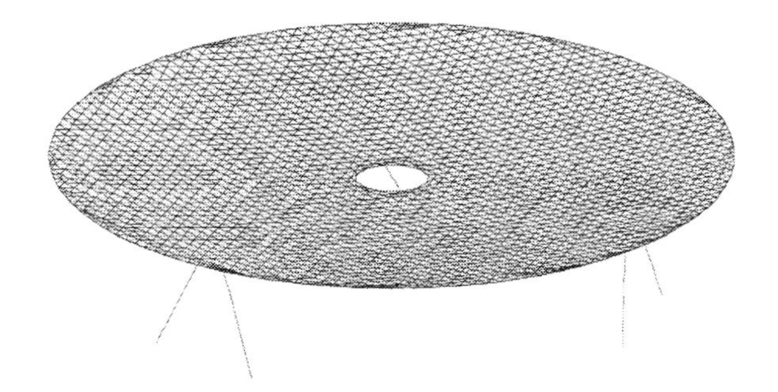

Figure 3.6: Two-dimensional equivalent-stiffness model of a lightweight mirror.

3.1.4.1 Two-dimensional equivalent-stiffness models of lightweight mirrors

In a 2D equivalent-stiffness model such as that shown in Fig. 3.6, effective plate properties are assigned to a plate mesh, representing the lightweight optic's construction. A ring of beam elements should also be included around the inner and outer edges of the mirror to represent the edge walls of the core. The properties of these beam elements should be computed with conventional beam-section equations. The grid plane of the 2D-element mesh may be placed at the neutral plane of the optic, or at any other convenient location, with the use of an offset definition. The definition of variables used in the equations for computing the effective properties are defined with Figs. 3.7 and 3.8 as follows:

t_f = front-faceplate thickness,
t_b = back-faceplate thickness,
t_c = core-wall thickness,
h_c = core height,
ρ = mass density, and
B = midplane-to-midplane inscribed circle cell size.

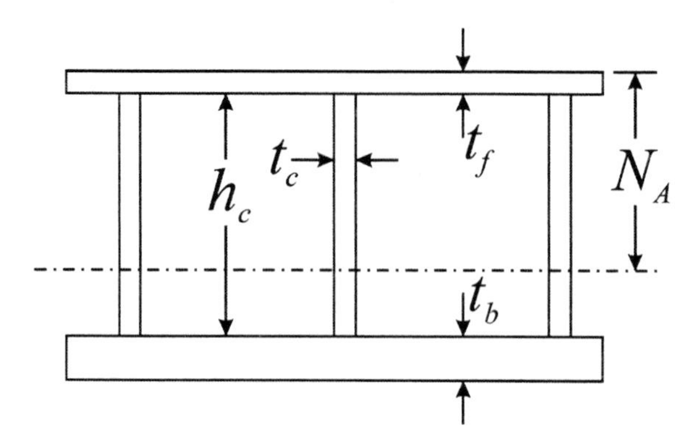

Figure 3.7 Variable definition for 2D effective model equations.

Figure 3.8 illustrates the definition of *B* for various cell shapes. If the mirror includes only an optical facesheet with an open-back triangular cell core, then the analyst is advised to use the distance between parallel core webs for *B* instead of the inscribed circle diameter used for closed-back mirrors. The rationale behind this method can be best illustrated by studying the bending deformation of the open back mirror shown in Fig. 3.9. Notice that the bending stiffness is dominated only by core walls, which are perpendicular to the moment axis. Core walls that are not oriented perpendicular to the axis of an applied plate-bending moment only twist around and do not significantly contribute to the stiffness of the bending section. Thus, the inscribed circle between parallel walls is chosen for open back mirrors.

The solidity ratio, α, is computed first from

$$\alpha = \frac{t_c}{B}. \tag{3.1}$$

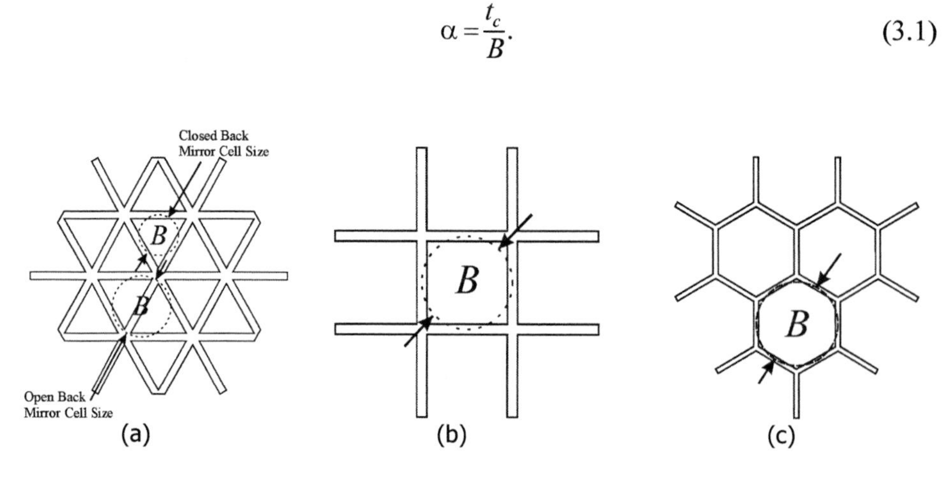

Figure 3.8 Cell size, *B*, definitions for various cell geometries: (a) triangular cells, (b) square cells, and (c) hexagonal cells. Notice the different cell size definitions for an open-back lightweight mirror vs. a closed-back lightweight mirror.

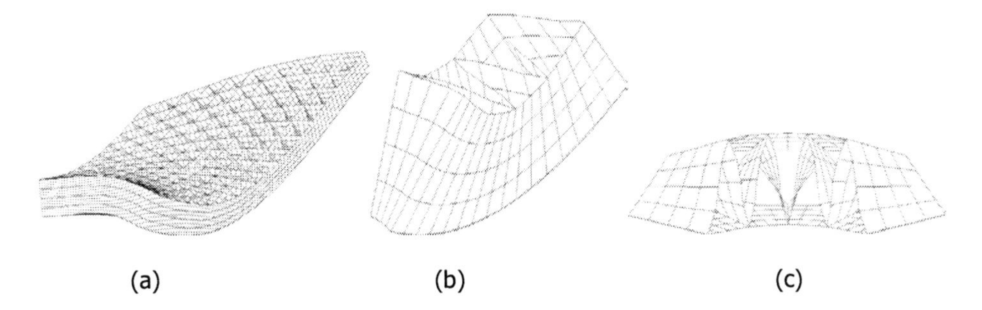

Figure 3.9 The core walls of an open back lightweight mirror display twisting deformation when the optic is loaded in bending; (a) an open back lightweight mirror in bending, (b) isometric view of some of the core cells, and (c) top view of some of the core cells showing twisting of core walls.

The effective membrane thickness, T_m, is computed by summing the front and back facesheet thicknesses with the core depth scaled by the solidity ratio.

$$T_m = t_f + t_b + \alpha h_c, \quad t_f \neq t_b, \qquad [3.2(a)]$$

or

$$T_m = 2t + \alpha h_c, \qquad t_f = t_b = t. \qquad [3.2(b)]$$

With the solidity ratio and other dimensions defined in Fig. 3.7, the distance of the neutral plane from the optical surface can be found from

$$N_A = \frac{1}{T_m}\left[t_f \frac{t_f}{2} + t_b\left(\frac{t_b}{2} + h_c + t_f\right) + \alpha h_c\left(\frac{h_c}{2} + t_f\right)\right], \quad t_f \neq t_b, \qquad [3.3(a)]$$

$$N_A = \frac{h_c}{2} + t, \qquad t_f = t_b = t. \qquad [3.3(b)]$$

The plate-bending moment of inertia is then computed from

$$I_b = \frac{1}{12}t_f^{\,3} + t_f\left(N_A - \frac{t_f}{2}\right)^2 + \frac{1}{12}t_b^{\,3} + t_b\left(N_A - t_f - h_c - \frac{t_b}{2}\right)^2 + \frac{1}{12}\alpha h_c^3 + \alpha h_c\left(N_A - t_f - \frac{h_c}{2}\right)^2, \quad t_f \neq t_b, \qquad [3.4(a)]$$

or

$$I_b = \frac{1}{12}\left[(2t + h_c)^3 - (1-\alpha)h_c^{\,3}\right], \qquad t_f = t_b = t. \qquad [3.4(b)]$$

Since some finite element codes require the bending moment of inertia be given as a scale factor on the quantity $T_m^3/12$, we can define a bending ratio, R_b, as

$$R_b = \frac{12 I_b}{T_m^3} . \tag{3.5}$$

The effective plate shear depth, S, can be found from

$$S = \frac{\alpha I_b}{\left(t_f + t_b + h_c\right)^2 - (1-\alpha) h_c^2}, \qquad t_f \neq t_b, \qquad [3.6(a)]$$

or

$$S = \frac{\alpha I_b}{\left(2t + h_c\right)^2 - (1-\alpha) h_c^2}, \qquad t_f = t_b = t. \qquad [3.6(b)]$$

As was done for the bending moment of inertia, a shear ratio can be expressed as

$$R_s = \frac{8S}{T_m} . \tag{3.7}$$

The effective membrane thickness, T_m, and the mass density, ρ, will not generate the correct mass representation in the 2D equivalent-stiffness model. Therefore, the model mass can be corrected for closed-back mirror models by adding nonstructural mass, NSM, defined as

$$\mathrm{NSM} = \rho \alpha h_c . \tag{3.8}$$

For open-back triangular core mirrors, the nonstructural mass must be twice the value computed by Eq. (3.8). The stress-recovery points as distances from the neutral plane are defined as

$$\begin{aligned} c_1 &= N_A \\ c_2 &= N_A - t_f - t_b - h_c, \end{aligned} \qquad t_f \neq t_b, \qquad [3.9(a)]$$

or

$$\begin{aligned} c_1 &= N_A \\ c_2 &= N_A - 2t - h_c, \end{aligned} \qquad t_f = t_b = t. \qquad [3.9(b)]$$

These equations assume that the element normals are directed from the back of the mirror toward the optical surface.

Despite the inability of 2D equivalent-stiffness models to predict quilting deformation, an estimation of the peak-to-valley of quilting can be independently computed by

$$\delta_{\text{Quilting}} = \frac{12\lambda p B^4 \left(1 - \nu^2\right)}{E t_f^{\,3}}, \tag{3.10}$$

where δ_{Quilting} is the peak-to-valley deformation, p is the applied pressure, and λ is a shape-dependent constant found in Table 3.1.[1] A surface root-mean-square (rms) value is found by scaling the peak-to-valley prediction by 0.3. This prediction can be combined to the model-predicted surface rms error by the root-sum-square (rss) method.

Table 3.1 Constants for use with Eq. (3.10).

CELL SHAPE	λ
Triangle	**0.00151**
Square	**0.00126**
Hexagon	**0.00111**

Figure 3.10 shows comparisons of the transverse displacement contributions from bending and transverse-shear compliances as a function of diameter-to-depth ratios, *(D/h)*, for a simply supported constant thickness mirror and a simply supported lightweight mirror with uniform pressures applied. Notice that since the fractional contribution of transverse-shear deformation does not become insignificant compared to the bending deformation in a lightweight mirror until diameter-to-depth ratios approach 100, it is extremely important to include an appropriate effective shear factor in order to develop an accurate representation of the mirror compliance.

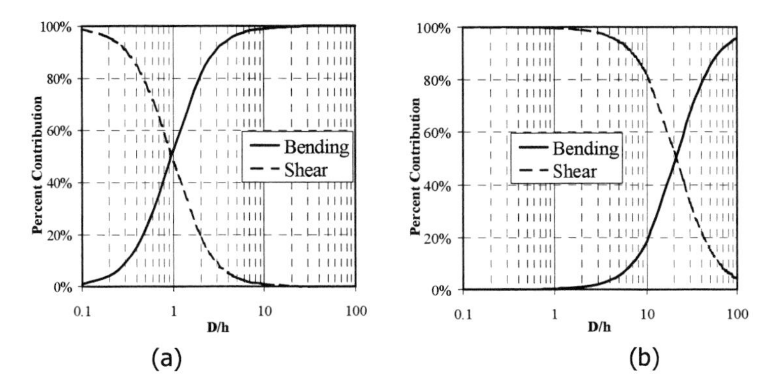

Figure 3.10 Bending- and shear-deformation contributions: (a) solid constant-thickness mirror and (b) lightweight mirror.

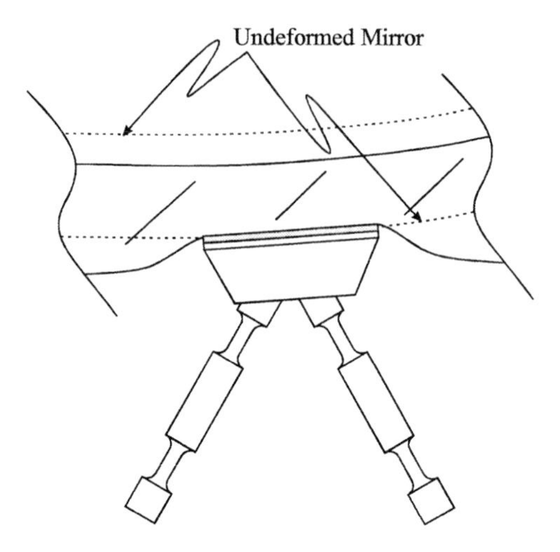

Figure 3.11 Highly exaggerated local deformation due to loads at the mounts.

The limitations of this model type are very similar to those discussed in Sec. 3.1.3.1 for 2D models of solid optics. In general, the global deformations of this type of model are reasonable for static and dynamic analyses. Most local effects such as mount dimpling are not well represented, and others such as quilting are not represented at all. In addition, since the stiffness through the depths of lightweight mirrors can be very low and their depths can be high compared to solid mirrors, the assumption that the through-the-thickness deformations are negligible may not be applicable for more strict analysis goals. For example, the axial optical surface rigid-body motion of a deep lightweight mirror subject to axial inertial loads may be very dependent on how much local deformation develops around the back surface mount points as shown in Fig. 3.11. A 2D-effective model lacks the ability to include these effects.

A unique advantage of the 2D equivalent-stiffness model is that it is easily implemented in a design optimization study. All of the effective property equations and the quilting estimate shown above may be included in a property sizing design optimization run to assist in the development of a lightweight mirror design to meet optical performance, weight, and other requirements. Although the accuracy of this model is not as favorable as the model types discussed below, it is the most superior model type for the purpose of quickly developing an optimum mirror design to be used later in more detailed verification analyses.

3.1.4.2 Three-dimensional equivalent-stiffness models

The 3D equivalent-stiffness model of a lightweight mirror, shown in Fig. 3.12, has predictive accuracy capabilities superior to the 2D equivalent-stiffness model but is slightly more complex. The optical and back faceplates are a mesh of plate elements that reside at the appropriate midsurfaces. They each reference unmodified material properties and the thickness of the faceplates. The lightweighted core, however, is represented by solid elements that share the nodes of the faceplate meshes and reference effective transversely-orthotropic material properties calculated from equations. In addition, as shown in Fig. 3.12, the 3D equivalent-stiffness model should also include a representation of the core edge wall with shell elements at the inner and outer mesh faces of the solid elements that represent the core.

The equations for computing the effective core properties are given below in two forms. The first set of equations gives the first order values of the engineering constants while the second set gives the elastic Hooke's law matrix, which relates the stresses to the strains. Both definitions are given to accomodate the requirements of different finite element codes.

The equations for the engineering constants are developed by the authors and are given as

$$\begin{aligned} &E_x^* = E_y^* = \alpha E, && E_z^* = 2\alpha E, \\ &\nu_{zx}^* = \nu_{zy}^* = \nu, && \nu_{xy}^* = \nu_{yx}^* = 0, \\ &\nu_{xz}^* = \nu_{yz}^* = \frac{\nu}{2}, && G_{xz}^* = G_{yz}^* = \alpha G, \\ &G_{xy}^* = 0, && \rho^* = \frac{2\alpha\rho h_c}{\frac{t_f}{2} + \frac{t_b}{2} + h_c}. \end{aligned} \tag{3.11}$$

where, E is the Young's modulus, G is the shear modulus, ν is the Poisson's ratio, ν_{ij} equals $-\varepsilon_j/\varepsilon_i$ due to a uniaxial stress applied in the i direction, ρ is the mass density, and * indicates an effective material property. Notice that the effective core density ρ^* includes a correction factor to account for the overlap in core mesh with half of each facesheet thickness.

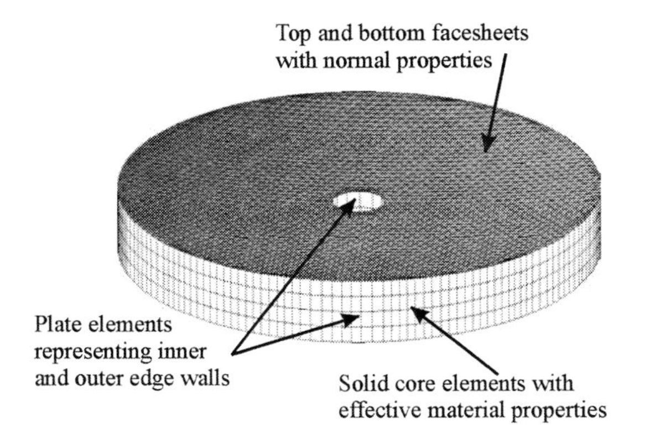

Figure 3.12 Three-dimensional equivalent-stiffness model of a lightweight mirror.

When Eq. (3.11) is substituted into the orthotropic form of Hooke's law as found in Jones,[2] the following matrix relation results:

$$\begin{Bmatrix} \sigma_{xx} \\ \sigma_{yy} \\ \sigma_{zz} \\ \tau_{xy} \\ \tau_{yz} \\ \tau_{zx} \end{Bmatrix} = \begin{bmatrix} \left(1-\frac{\nu^2}{2}\right)\frac{\alpha E}{\left(1-\nu^2\right)} & \frac{\nu^2\alpha E}{2\left(1-\nu^2\right)} & \frac{\nu\alpha E}{\left(1-\nu^2\right)} & 0 & 0 & 0 \\ \frac{\nu^2\alpha E}{2\left(1-\nu^2\right)} & \left(1-\frac{\nu^2}{2}\right)\frac{\alpha E}{\left(1-\nu^2\right)} & \frac{\nu\alpha E}{\left(1-\nu^2\right)} & 0 & 0 & 0 \\ \frac{\nu\alpha E}{\left(1-\nu^2\right)} & \frac{\nu\alpha E}{\left(1-\nu^2\right)} & \frac{2\alpha E}{\left(1-\nu^2\right)} & 0 & 0 & 0 \\ 0 & 0 & 0 & 0 & 0 & 0 \\ 0 & 0 & 0 & 0 & \frac{\alpha E}{2\left(1+\nu\right)} & 0 \\ 0 & 0 & 0 & 0 & 0 & \frac{\alpha E}{2\left(1+\nu\right)} \end{bmatrix} \bullet \begin{Bmatrix} \varepsilon_{xx} \\ \varepsilon_{yy} \\ \varepsilon_{zz} \\ \gamma_{xy} \\ \gamma_{yz} \\ \gamma_{zx} \end{Bmatrix}. \tag{3.12}$$

The reader should notice the ordering of the elements of the stress and strain vectors in Eq. (3.12). The order of these elements varies throughout the literature.

Since the effective material properties of the core are dependent on direction, it is important for the analyst to make sure that the material coordinate system of the solid-element mesh is correctly defined so that the material description will be properly oriented. Since the x and y directions are identical in the above formulations, either a cylindrical or rectangular material coordinate system may be employed as long as the z direction is defined parallel to the direction defined by the intersection of the core walls.

The 3D equivalent-stiffness model predicts some deformation behaviors not represented in the 2D equivalent-stiffness model but still displays some shortcomings in predictive accuracy. Global-elastic behavior through the thickness of the mirror is well represented. Deformation effects such as thermoelastic growth through the thickness and elastic isolation of the optical surface from the mount points is represented quite well. However, highly localized effects at the mount points are not fully represented. Therefore, while the optical surface rigid-body motion is better predicted with a 3D equivalent-stiffness model compared to the 2D equivalent-stiffness model, some inaccuracies are, nevertheless, to be expected. In addition, quilting deformation is not represented at all. Eq. (3.10) can be employed to estimate quilting effects as was suggested for the 2D equivalent-stiffness model.

The 3D equivalent-stiffness model has many of the same benefits of simplicity as the 2D equivalent-stiffness model, but it has increased predictive capability. Its use in design optimization, however, requires features that allow the analyst to define material properties as design variables. In addition, the consideration of mirror depth as a design variable requires shape optimization features not available in all codes. Nevertheless, some codes are available that contain both of these features. Such capabilities make the 3D equivalent-stiffness model an excellent choice for preliminary design trade studies where through-the-thickness effects may be very important.

3.1.4.3 Three-dimensional plate/shell model

The 3D plate/shell model has the most superior deformation prediction capabilities, but it is the most complicated and time consuming model type to construct. It is also the most difficult model type to alter, often making it a poor choice for early design optimization trade studies. The model is composed entirely of plate elements located at the midsurfaces of each facesheet and core wall segment. An example model is shown in Fig. 3.13. Effective properties can be given to the facesheets using the 2D equivalent-stiffness method if the facesheets contain their own cathedral-rib stiffening.

Due to the geometry of most lightweight mirror cores, it is unlikely that the analyst will be able to mesh the mirror faceplates of quadrilateral elements without some degree of warping. Since warping is an extremely detrimental distortion for four-noded quadrilateral elements, it is advised that three-noded triangular elements be employed. Some finite element codes have developed three-noded elements, which have formulations superior to a constant strain formulations, but such elements are not always fully featured. The three-noded constant-strain formulation elements give excellent results if an adequate number of elements are used.

As was done in the 3D equivalent-stiffness model, the mass density of the core elements can be adjusted to account for the overlap of the core elements with half of each facesheet thickness. This adjustment is performed by scaling the true mass density by $h_c/(t_f/2 + t_b/2 + h_c)$.

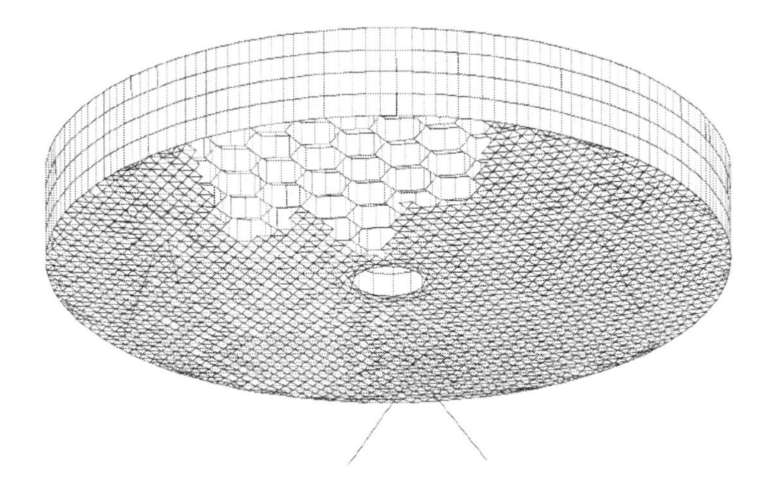

Figure 3.13 Three-dimensional plate/shell model of a lightweight mirror.

3.1.4.4 Example: gravity deformation prediction comparison of a lightweight mirror

Predictions of natural frequencies, weight, and static deformation due to a gravity load from each of the three model types discussed above are to be compared for a lightweight mirror design fabricated of ULE.[3] The mirror has an outer diameter of 71.12 cm and an inner diameter of 7.62 cm. The core depth is 5 cm, and the cells are hexagonal in shape with a midplane-to-midplane inscribed circle diameter of 5 cm. The facesheets are 4.6-mm thick and the core-wall thickness is 1.5 mm. The mirror is mounted on three sets of bipod flexures that are bonded to the back surface of the optic.

3.1.4.4(a) Two-dimensional effective property calculations

By Eq. (3.1) we compute the solidity ratio:

$$\alpha = \frac{t_c}{B} = \frac{1.5\ \text{mm}}{50\ \text{mm}} = 0.03\,.$$

The effective-membrane thickness, T_m, is computed by Eq. [3.2(b)]:

$$T_m = 2t + \alpha h_c$$

$$T_m = 2(4.6\ \text{mm}) + (0.03)(50.\ \text{mm}) = 10.7\ \text{mm}\,.$$

Find the location of the neutral plane, N_A, using Eq. [3.3(b)]:

$$N_A = \frac{h_c}{2} + t\ ,$$

$$N_A = \frac{50.\ \text{mm}}{2} + 4.6\ \text{mm} = 29.6\ \text{mm}\,.$$

The plate-bending moment of inertia is computed with Eq. [3.4(b)]:

$$I_b = \frac{1}{12}\left[(2t + h_c)^3 - (1-\alpha)h_c^{\,3}\right]$$

$$I_b = \frac{1}{12}\left\{\left[2(4.6\ \text{mm}) + (50.\ \text{mm})\right]^3 - \left[1-(0.03)\right](50.\ \text{mm})^3\right\}$$

$$= \frac{1}{12}\left[207{,}474.688\ \text{mm}^3 - 121{,}250.0\ \text{mm}^3\right]$$

$$= 7{,}185.39\ \text{mm}^3.$$

The bending ratio, I_b, is computed by Eq. (3.5):

$$R_b = \frac{12 I_b}{T_m^3},$$

$$R_b = \frac{12\left(7185.39\ \text{mm}^3\right)}{\left(10.7\ \text{mm}\right)^3} = 70.385\,.$$

The effective shear depth, S, is computed from Eq. [3.6(b)]:

$$S = \frac{\alpha I_b}{\left(2t + h_c\right)^2 - \left(1-\alpha\right)h_c^2}$$

$$S = \frac{(0.03)\left(7185.39\ \text{mm}^3\right)}{\left[2(4.6\ \text{mm}) + 50.0\ \text{mm}\right]^2 - (1-0.03)(50.0\ \text{mm})^2} = 0.200\ \text{mm}\,.$$

By Eq. (3.7) the shear-factor ratio is then

$$R_s = \frac{8S}{T_m} = \frac{8(0.200\ \text{mm})}{(10.7\ \text{mm})} = 0.149.$$

Equation (3.8) is used to compute the nonstructural mass that corrects the model mass with a material density of 2.187 g/cm^3 as

$$\text{NSM} = \rho\alpha h_c = \left(0.002187\text{g}/\text{mm}^3\right)(0.03)(50.0\text{mm}) = 0.00328\text{g}/\text{mm}^2$$

The 2D equivalent-stiffness model is shown in Fig. 3.6.

3.1.4.4(b) Three-dimensional effective property calculations

Since the facesheets are modeled with their true thicknesses, effective properties for the 3D equivalent-stiffness model are computed only for the core. As shown in Eq. (3.12) these properties are in the form of a Hooke's law matrix whose elements are G_{ij} for the ith row and the jth column. The computations for the nonzero values of the Hooke's law matrix are shown as

$$G_{11} = G_{22} = \left(1 - \frac{\nu^2}{2}\right)\frac{\alpha E}{1-\nu^2} = \left[1 - \frac{(0.17)^2}{2}\right]\frac{(0.03)\left(6.757\times 10^{10}\,\mu N/\mathrm{mm}^2\right)}{1-(0.17)^2}$$
$$= 2.057\times 10^{9}\,\mu N/\mathrm{mm}^2,$$

$$G_{12} = G_{21} = \left(\frac{\nu^2}{2}\right)\frac{\alpha E}{1-\nu^2} = \left[\frac{(0.17)^2}{2}\right]\frac{(0.03)\left(6.757\times 10^{10}\,\mu N/\mathrm{mm}^2\right)}{1-(0.17)^2}$$
$$= 3.016\times 10^{7}\,\mu N/\mathrm{mm}^2,$$

$$G_{13} = G_{23} = G_{31} = G_{32} = \nu\frac{\alpha E}{1-\nu^2} = (0.17)\frac{(0.03)\left(6.757\times 10^{10}\,\mu N/\mathrm{mm}^2\right)}{1-(0.17)^2}$$
$$= 3.549\times 10^{8}\,\mu N/\mathrm{mm}^2,$$

$$G_{33} = 2\frac{\alpha E}{1-\nu^2} = 2\frac{(0.03)\left(6.757\times 10^{10}\,\mu N/\mathrm{mm}^2\right)}{1-(0.17)^2} = 4.175\times 10^{9}\,\mu N/\mathrm{mm}^2,$$

$$G_{55} = G_{66} = \frac{\alpha E}{2(1+\nu)} = \frac{(0.03)\left(6.757\times 10^{10}\,\mu N/\mathrm{mm}^2\right)}{2(1+0.17)} = 8.663\times 10^{8}\,\mu N/\mathrm{mm}^2.$$

The effective core density is given by Eq. (3.11).

$$\rho^* = \frac{2\alpha\rho h_c}{\frac{t_f}{2} + \frac{t_b}{2} + h_c}$$
$$= \frac{2(0.03)\left(0.002187\,\mathrm{g/mm}^3\right)(50.0\,\mathrm{mm})}{\frac{4.6\,\mathrm{mm}}{2} + \frac{4.6\,\mathrm{mm}}{2} + 50.0\mathrm{mm}} = 1.202\times 10^{-4}\,\mathrm{g/mm}^3.$$

The 3D equivalent-stiffness model is shown in Fig. 3.12.

3.1.4.4(c) Three-dimensional plate/shell model effective property calculations

The only effective property to compute for the 3D plate/shell model is the core density. Since the mesh of the core extends through half of the dimension of the faceplates, the nominal density is scaled as follows:

$$\rho^* = \rho \frac{h_c}{\frac{t_f}{2} + \frac{t_b}{2} + h_c}$$

$$= \left(0.002187\text{g}/\text{mm}^3\right) \frac{50.0\text{ mm}}{\frac{4.6\text{ mm}}{2} + \frac{4.6\text{ mm}}{2} + 50.0\text{ mm}} = 0.002003\text{g}/\text{mm}^3.$$

The 3D plate/shell model is shown in Fig. 3.13.

3.1.4.4(d) Comparison of results

A comparison of the deformation results due to gravity is shown in Table 3.2. The deformation results have been formatted in nonzero Zernike polynomial coefficients with units of nanometers. Plots of the deformations for each model are shown in Fig. 3.14

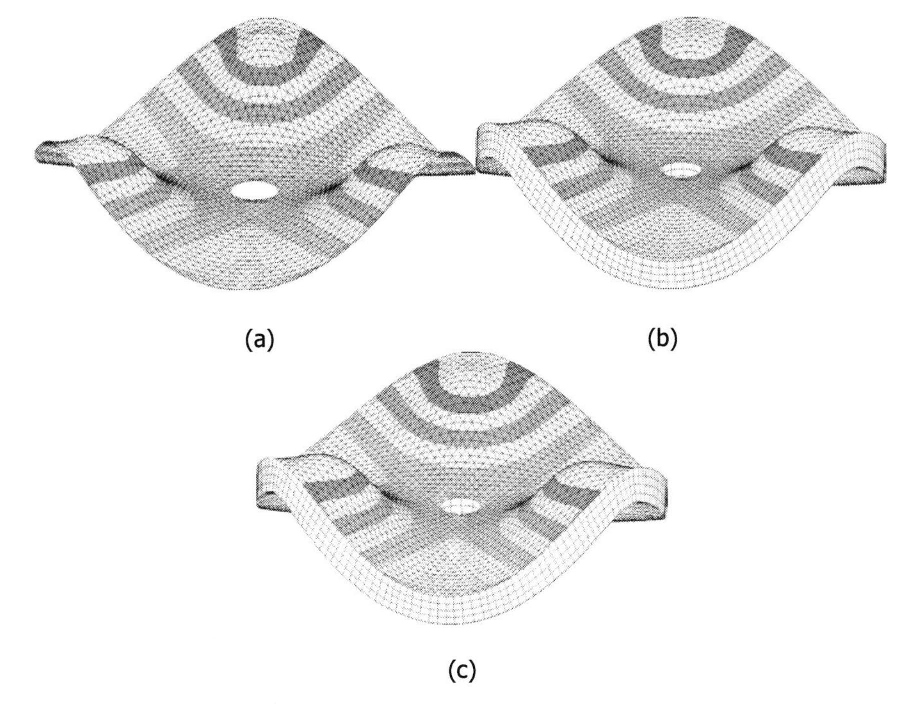

Figure 3.14 Highly exaggerated deformed plots of mounted lightweight mirror models loaded by gravity: (a) 2D equivalent-stiffness model, (b) 3D equivalent-stiffness model, and (c) 3D plate/shell model.

Table 3.2 Gravity deformation results.

	Mag. (nm)	Residual rms (nm)	Residual P-V (nm)	Mag. (nm)	Residual rms (nm)	Residual P-V (nm)	Mag. (nm)	Residual rms (nm)	Residual P-V (nm)
Input Surface		934.5	433.7		942.3	414.4		945.4	422.0
Bias	–928.6	112.4	433.7	–937.0	108.6	414.4	–939.8	110.8	422.0
Power (Defocus)	37.5	110.5	459.7	34.4	107.0	440.8	36.8	109.0	450.2
Pri Trefoil	287.7	41.9	220.3	278.4	40.7	218.4	283.1	41.9	227.1
Pri Spherical	–44.5	37.1	162.3	–42.9	36.2	160.2	–44.0	37.2	166.1
Sec Trefoil	112.0	17.7	99.6	110.8	16.3	94.3	113.8	16.9	97.4
Sec Spherical	8.2	17.3	87.7	10.3	15.8	79.5	10.7	16.4	82.1
Pri Hexafoil	39.1	14.0	90.7	34.9	12.9	88.3	36.0	13.4	91.8
Ter Trefoil	22.8	12.7	61.0	25.2	11.1	57.2	25.8	11.6	60.6
Ter Spherical	8.8	12.3	67.5	7.0	10.9	61.5	7.5	11.3	65.3
Sec Hexafoil	38.9	7.8	44.2	35.0	6.7	38.9	36.3	7.0	41.3

Notice that all three models yield similar displacement predictions, but the differences in the predictions and the plotted deformations illustrate the limitations of each. The inability of the 2D equivalent-stiffness model to represent through-the-thickness deformation is illustrated by a greater trefoil prediction and by a lower bias prediction as compared to the 3D equivalent-stiffness and 3D plate/shell models. The global deformation predictions are very similar, however, as evidenced by comparable residual rms predictions after bias is removed.

The weight and natural frequency predictions are shown in Table 3.3.

Table 3.3 Weight and natural frequency predictions.

	2D Effective	3D Effective	3D Plate
Weight	10.94 kg	10.96 kg	10.86 kg
Unmounted Natural Frequency	813 Hz	812 Hz	809 Hz
Mounted Natural Frequency	129 Hz	131 Hz	131 Hz

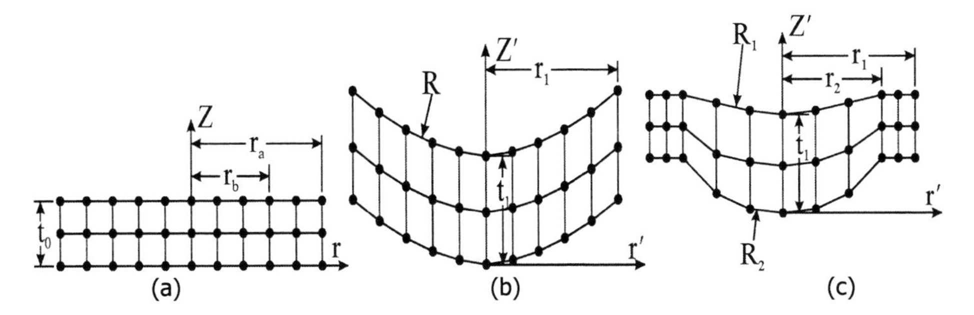

Figure 3.15 Definition of variables for powered-optic-model-generation equations: (a) initial model shape, (b) final mirror-model shape, and (c) final lens-model shape.

3.1.5 Generation of powered optic models

Often it is easiest to construct a finite element model of a lens or mirror as a flat optic, and then use a program or spreadsheet to modify the node coordinate values to obtain the final shape. Equations (3.13) and (3.14) are two example transformation relations that give the new coordinate values in terms of the flat optic coordinate values. The equations are expressed in polar coordinate variables and the variable definitions are given in Fig. 3.15. Figure 3.15(a) represents a flat model of a lens or mirror, which is to be transformed into either the mirror model shown in Fig. 3.15(b) or the lens model shown in Fig. 3.15(c). The transformation from Fig. 3.15(a) to Fig. 3.15(b) is as follows:

$$\begin{aligned} r' &= \frac{r_1}{r_a} r \\ Z' &= Z\frac{t_1}{t_0} + R - \sqrt{R^2 - r'^2}, \end{aligned} \tag{3.13}$$

while the transformation from Fig. 3.15(a) to Fig. 3.15(c) is as follows:

$$r' = \begin{cases} \dfrac{r_2}{r_b} r & 0 \le r \le r_b \\ r_2 + (r_1 - r_2)\dfrac{(r - r_b)}{(r_a - r_b)} & r_b \le r \le r_a \end{cases}$$

$$Z' = \begin{cases} Z\dfrac{t}{t_0} + \left(R_2 - \sqrt{R_2^2 - r'^2}\right)\left(\dfrac{t_0 - Z}{t_0}\right) + \left(R_1 - \sqrt{R_1^2 - r'^2}\right)\left(\dfrac{Z}{t_0}\right) & 0 \le r' \le r_2 \\ \left(R_2 - \sqrt{R_2^2 - r_2^2}\right)\left(\dfrac{t_0 - Z}{t_0}\right) + \left(R_1 - \sqrt{R_1^2 - r_2^2}\right)\left(\dfrac{Z}{t_0}\right) & r_2 \le r' \le r_1 \end{cases} . \tag{3.14}$$

Notice that the above methods can be applied to any of the optic displacement models discussed in this section.

3.2 Displacement Models of Adhesive Bonds

Adhesives commonly used to bond optics to their mounts are not trivial items to model. Characteristics such as near incompressibility and extremely small thicknesses pose difficulties to developing accurate numerical models for these bonds. However, low stiffness, high cure shrinkage, and high thermoelastic growth characteristics of adhesives require that they be well represented in optomechanical models in order to obtain useful predictions. Treatment of the effective modulus through-the-thickness of thin nearly incompressible bonds has been the topic of several sources in the published literature.[4,5,6] However, these sources have two principal limitations. None give a complete description of how the full elastic description of Hooke's law should be represented by effective properties for coarse bond models. Furthermore, no practical methods are presented by which one can obtain effective properties for various bond geometries. This section presents example methods of modeling adhesive bonds that accomplish both of these goals.[7]

3.2.1 Elastic behavior of adhesives

To familiarize the reader with the relevant aspects of adhesive-bond behavior we will first introduce two extreme cases of adhesive test samples. These cases shown in Fig. 3.16 are the thin uniaxial test sample and the thin-layer test sample. The material's incompressibility and the difference in geometries cause these two samples to behave with very different stress-to-strain ratios. While the uniaxial test sample freely allows the lateral strains required to allow straining in the loaded direction, the thin-layer test sample strongly resists such lateral strains. Therefore, the thin-layer test sample appears to behave with a higher stress-to-strain ratio as the lateral straining occurs only near the free surface of the bond. This behavior causes difficulties in modeling bonds, which are similar to the thin-layer test sample. In order to correctly represent the compliance of the bond, the local deformations near the free edge must be included. This section illustrates methods by which such behavior can be accurately represented in a finite element model.

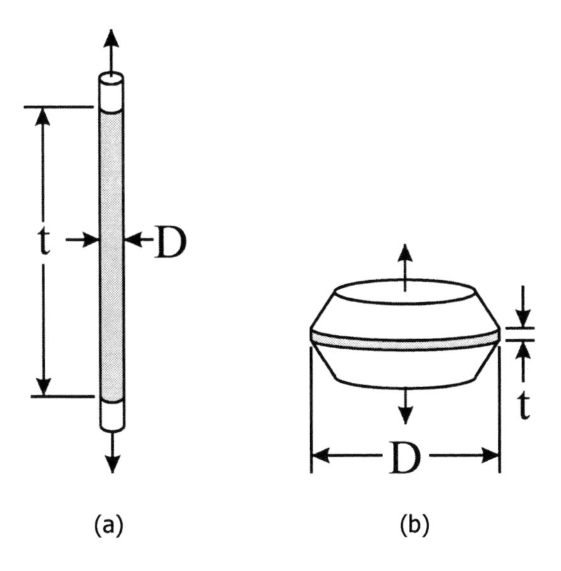

Figure 3.16 Uniaxial test sample and thin-layer test sample.

We can bound the stress-to-strain ratios of these two test samples by making assumptions about the stresses and strains in each case and applying them to the stress-to-strain relationships in Eqs. (1.2) and (1.3). For each of the two test samples a load is applied in the manner shown in Fig. 3.16 and the strain along the load direction, e_z, is calculated. From Hooke's law the expression for this strain in terms of the stresses is

$$e_z = -\frac{\nu}{E}\sigma_x - \frac{\nu}{E}\sigma_y + \frac{1}{E}\sigma_z. \tag{3.15}$$

For the uniaxial test sample shown in Fig. 3.16(a), we may assume that σ_x and σ_y are zero. This gives

$$\frac{\sigma_z}{e_z} = E, \tag{3.16}$$

which is the familiar uniaxial stress-strain relationship.

From Hooke's law the expression of the test load stress in terms of the strains is

$$\sigma_z = \frac{\nu E}{(1+\nu)(1-2\nu)}e_x + \frac{\nu E}{(1+\nu)(1-2\nu)}e_y + \frac{(1-\nu)E}{(1+\nu)(1-2\nu)}e_z. \tag{3.17}$$

For the thin-layer test shown in Fig. 3.16(b) we assume that e_x and e_y are 0. This gives

$$\frac{\sigma_z}{e_z} = \frac{(1-\nu)E}{(1+\nu)(1-2\nu)} = M, \tag{3.18}$$

which is defined as the *maximum modulus, M.* Notice from Eq. (3.18) that the maximum modulus is increasingly dependent on Poisson's ratios greater than about 0.45, and is undefined at a Poisson's ratio of 0.5. This dependence on Poisson's ratio is shown in Fig. 3.17 and Table 3.4. Notice that for Poisson's ratios greater than 0.49 each additional "9" adds an order of magnitude to the maximum modulus. Recall that if Poisson's ratio equals 0.5, then the material is incompressible.

We can now imagine the spectrum of cases bounded by the uniaxial and thin layer extremes discussed above. These test samples each have a unique diameter-to-thickness ratio, D/t, where D is the diameter of the test sample perpendicular to the applied load, and t is the thickness parallel to the applied load. With a series of detailed finite element analyses, the ratio of applied stress to axial modulus, E, is shown in Fig. 3.18(a), while the comparison to the maximum modulus, M, is shown in Fig. 3.18(b).

Table 3.4 Maximum modulus to Young's modulus ratio vs. Poisson's ratio.

POISSON'S RATIO	M/E
0.45	3.8
0.49	17.1
0.499	167.1
0.4999	1667.1
0.49999	16667.1

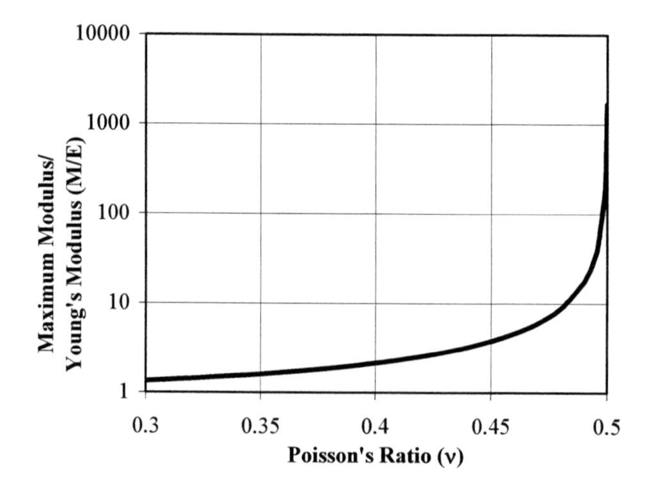

Figure 3.17 Plot of maximum modulus, M, divided by Young's modulus, E, vs. Poisson's ratio, ν.strain, (σ/ε), can be predicted for each D/t ratio.

These computed stress-to-strain ratios can be compared to the stress-to-strain ratios for each of the extreme cases as computed with Eqs. (3.16) and (3.18). The comparison to the uniaxial modulus, E, is shown in Fig. 3.18(a), while the comparison to the maximum modulus, M, is shown in Fig. 3.18(b).

As would be expected, the comparison to the uniaxial modulus is closer for more uniaxial cases than for thin-layer cases. In other words, $(\sigma/\varepsilon)/E$ in Fig. 3.18(a) approaches unity for cases with low diameter-to-thickness ratios. Likewise, $(\sigma/\varepsilon)/M$ in Fig. 3.18(b) approaches unity for cases with higher diameter-to-thickness ratios. Between the two extremes are cases that possess complex strain states. These complex strain states are characterized by the radial-edge deformation that can contribute to a large percentage of the compliance of the bond (see Fig. 3.19).

In addition to the dependence on a diameter-to-thickness ratio, the overall stiffness varies with Poisson's ratio. Higher values of Poisson's ratio show less agreement with either of the two extremes for a given diameter-to-thickness ratio, because higher Poisson's ratios weaken the validity of the assumptions used to generate Eqs. (3.16) and (3.18). Therefore, larger values of Poisson's ratio yield a

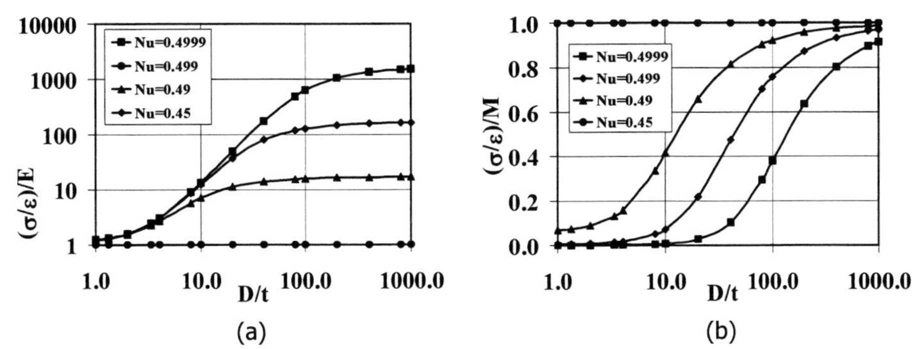

Figure 3.18 (a) Plot of stress-to-strain ratio, (σ/ε), divided by Young's modulus, E, vs. diameter-to-thickness ratio, D/t, and (b) plot of stress-to-strain ratio, (σ/ε), divided by maximum modulus, M, vs. diameter-to-thickness ratio, D/t.

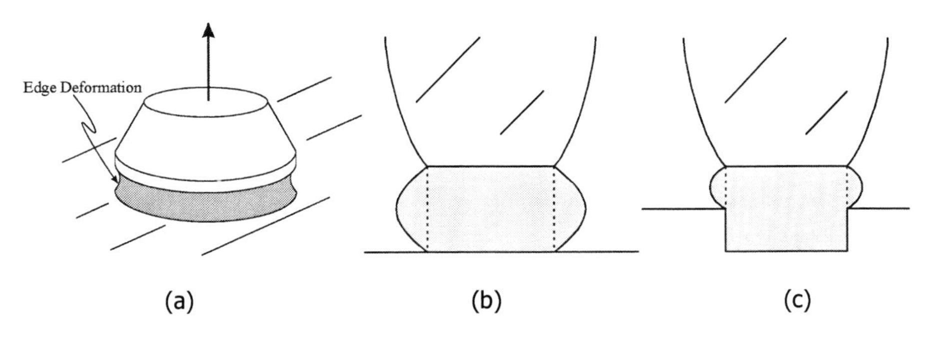

Figure 3.19 Exaggerated illustrations of incompressible bond behavior: (a) "hockey-puck" type bond (b) ring bond of a lens (c) partially constrained ring bond of a lens.

wider range of diameter-to-thickness ratios that do not behave like either of the two extreme cases presented above.

The objective of the foregoing discussion is to illustrate two important aspects when modeling nearly incompressible bonds. The first is that it is essential that an accurate value of Poisson's ratio be included in the model since the stiffness of nearly incompressible bonds is highly sensitive to this parameter. The second aspect is that the finite element mesh must be capable of predicting the free-face deformations such as those illustrated by Fig. 3.19.

3.2.2 Detailed three-dimensional solid model

One obvious method of modeling adhesive bonds is to use solid elements with enough resolution to represent their nearly incompressible behavior. Four elements or more should be used along free surfaces to represent the deformation effects illustrated in Fig. 3.19. Enough elements should be used in the plane of the bond to represent the decay in the edge deformation as well. The Young's modulus, E, and bulk modulus, B, can be obtained from tests. Poisson's ratio can then be calculated from the Young's modulus and bulk modulus:

$$\nu = \frac{1}{2} - \frac{E}{6B}. \tag{3.19}$$

The shear modulus can then be computed as

$$G = \frac{E}{2(1+\nu)}. \tag{3.20}$$

3.2.3 Equivalent stiffness bond models

The disadvantage of the detailed 3D solid model is that the fidelity required in order to properly represent the stiffness of the bond is often too fine to be practically integrated with the adjacent displacement models of the bonded parts. In addition, when several elements are used through the thickness of the bond, unacceptable aspect ratios can result. Therefore, it is of interest to obtain a method of using coarse meshes of adhesive bonds without sacrificing an accurate representation of the stiffness of the bond. The method involves using effective properties with a 3D solid model of the adhesive bond that is coarse enough such that the free-face deformation is not represented at all. The effective properties, however, are chosen such that the compliance of the bond due to the free-face deformation is included in the coarse model's stiffness.

Calculation of the effective properties involves modeling the adhesive bond in a detailed test model and computing the bond's overall stiffness. From these stiffness predictions, an effective Hooke's law matrix relating stress to strain can be computed for use as effective properties in a coarse model. Fortunately, these effective properties are functions of the uniaxial test material properties and a small number of geometric parameters. Therefore, effective property curves can be generated as functions of these parameters to create "look-up" tables. The relevant geometric parameters vary, however, for different applications.

3.2.3.1 Effective properties for "hockey-puck" type bonds

A "hockey-puck" bond is a thin, relatively flat bond such as that illustrated in Fig. 3.20. The bond may be circular or a more complex shape. The method for computing the effective properties of "hockey-puck" type bonds is as follows:

≺1≻ Compute the diameter-to-thickness ratio, *D/t*. For noncircular geometries an effective diameter, D_{eff}, is suggested by Lindley[3] and is computed as follows:

$$D_{\text{eff}} = \frac{4A}{C}, \tag{3.21}$$

where *A* is the plane-view area of the bond and *C* is the circumference of the plan-view area. Compute the diameter-to-thickness ratio using D_{eff} for the diameter.

≺2≻ Compute the maximum modulus, *M*, given by Eq. (3.18), with values of *E* and ν.

≺3≻ With ν and *D/t*, find the correction factors, k_{33} and k_{31}, from Table 3.5.

≺4≻ Use one of the modeling methods described in the text below.

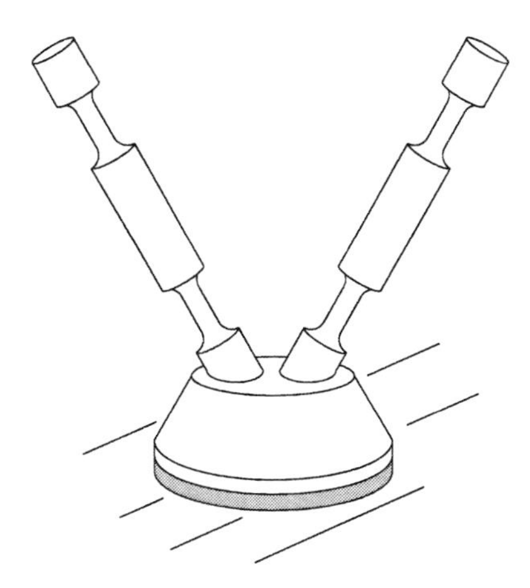

Figure 3.20 Example of a hockey-puck bond.

Table 3.5 Correction factors for "hockey puck" bonds with various combinations of *D/t* ratio and Poisson's ratio.

	ν = 0.45		ν = 0.49		ν = 0.499		ν = 0.4999	
***D/t* Ratio**	k_{33}	k_{31}	k_{33}	k_{31}	k_{33}	k_{31}	k_{33}	k_{31}
1	0.3069	0.1973	0.0710	0.1918	0.0073	0.1908	0.0007	0.1907
2	0.3665	0.3862	0.0900	0.3761	0.0095	0.3741	0.0010	0.3739
5	0.5804	0.7443	0.2014	0.7555	0.0244	0.7604	0.0025	0.7609
10	0.7624	0.8908	0.4172	0.9141	0.0739	0.9250	0.0080	0.9263
20	0.8746	0.9507	0.6579	0.9682	0.2198	0.9788	0.0295	0.9803
50	0.9458	0.9814	0.8505	0.9895	0.5580	0.9953	0.1508	0.9966
100	0.9700	0.9908	0.9209	0.9950	0.7574	0.9981	0.3797	0.9990
200	0.9822	0.9954	0.9573	0.9976	0.8715	0.9991	0.6342	0.9997
500	0.9897	0.9981	0.9794	0.9990	0.9440	0.9997	0.8394	0.9999
1000	0.9927	0.9992	0.9869	0.9995	0.9689	0.9998	0.9151	0.9999

Figures 3.21 and 3.22 show plots of k_{33} and k_{31} vs. *D/t* ratio for the values of Poisson's ratio shown in Table 3.5. However, it is advised that values be taken by interpolation from Table 3.5 rather than graphically from Figures 3.21 and 3.22.

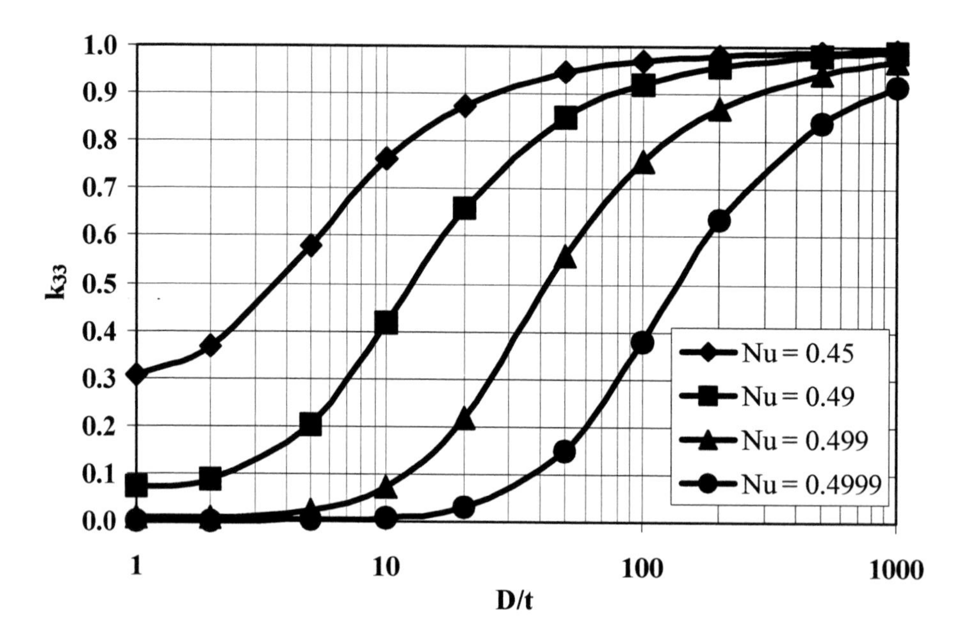

Figure 3.21 Plots of k_{33} vs. *D/t* for various values of Poisson's ratio.

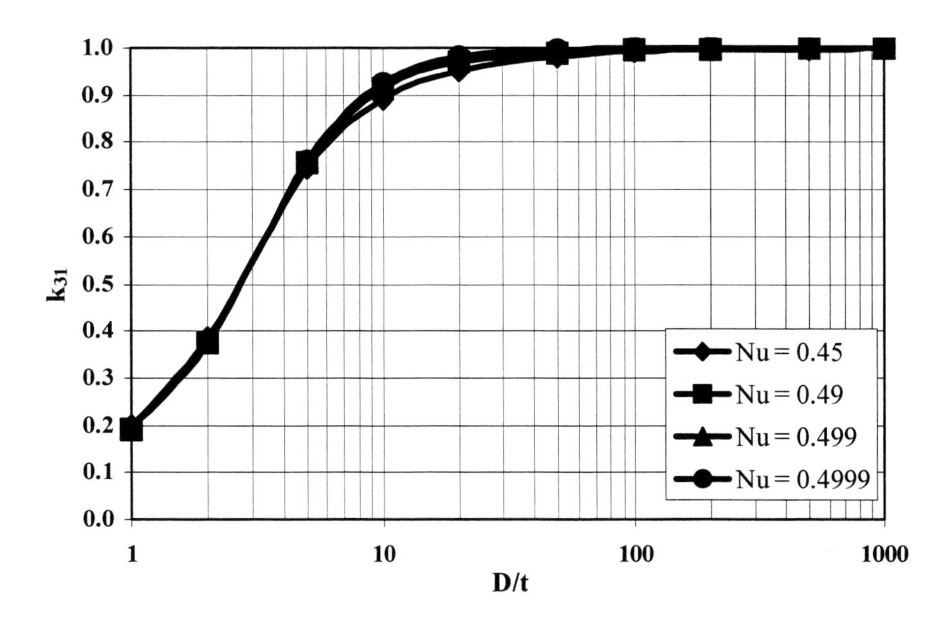

Figure 3.22 Plots of k_{31} vs. D/t for various values of Poisson's ratio.

There are several methods of using the correction factors, k_{33} and k_{31}, to obtain effective properties. One method is to mesh the adhesive bond with solid elements and use only one element through the thickness. A mesh fidelity in the plane of the bond could be chosen to reasonably match the mesh of the models being connected. An effective form of Hooke's law for the coarse adhesive-bond model is defined in Eq. (3.22). Eq. (3.22) assumes the "3" direction is through the thickness of the bond while the "1" and "2" directions are in the plane of the bond. The analyst should be careful to orient the material coordinate system of the adhesive mesh such that the material description in Eq. (3.22) is aligned correctly with respect to the through-the-thickness direction. The rows and columns of the Hooke's law matrix may be rearranged to facilitate this:

$$\begin{Bmatrix} \sigma_{11} \\ \sigma_{22} \\ \sigma_{33} \\ \tau_{12} \\ \tau_{23} \\ \tau_{31} \end{Bmatrix} = \begin{bmatrix} M & \dfrac{\nu M}{(1-\nu)} & \dfrac{k_{31}k_{33}\nu M}{(1-\nu)} & 0 & 0 & 0 \\ \dfrac{\nu M}{(1-\nu)} & M & \dfrac{k_{31}k_{33}\nu M}{(1-\nu)} & 0 & 0 & 0 \\ \dfrac{k_{31}k_{33}\nu M}{(1-\nu)} & \dfrac{k_{31}k_{33}\nu M}{(1-\nu)} & k_{33}M & 0 & 0 & 0 \\ 0 & 0 & 0 & G & 0 & 0 \\ 0 & 0 & 0 & 0 & G & 0 \\ 0 & 0 & 0 & 0 & 0 & G \end{bmatrix} \begin{Bmatrix} \varepsilon_{11} \\ \varepsilon_{22} \\ \varepsilon_{33} \\ \gamma_{12} \\ \gamma_{23} \\ \gamma_{31} \end{Bmatrix} . \quad (3.22)$$

Alternatively, a beam element may be used to represent a hockey-puck bond. Such an approach might be used for single-point optic models or for very coarse models where a single node is to represent the bonded area. The mesh of the adhesive bond would be a single-beam element whose axis is oriented in the through-the-thickness direction. The geometric properties to be used for such a model are identical to the usual calculation of beam properties, where the cross-sectional area is the plan-view area of the bond and the Young's modulus, E, should be replaced by $k_{33}M$.

A third method of modeling a hockey-puck bond with the effective material properties is to use six scalar elastic elements or springs. The six-spring constants can be computed from

$$\begin{aligned} k_x &= k_y = \frac{KGA}{t} \\ k_z &= \frac{k_{33}MA}{t} \\ k_{\theta x} &= k_{\theta y} = \frac{k_{33}MI}{t} \\ k_{\theta z} &= \frac{GJ}{t}, \end{aligned} \tag{3.23}$$

where, $k_x = k_y$ are the in-plane shear stiffnesses of the bond, k_z is the through the thickness stiffness of the bond, $k_{\theta x} = k_{\theta y}$ are the rotational stiffnesses of the bond about the axes in the plane of the bond, and $k_{\theta z}$ is the torsional stiffness of the bond about the through-the-thickness direction.

3.2.3.2 Example: modeling of a hockey-puck-type bond

The bond in Fig. 3.20 is to be included in a finite element analysis of a mirror. The diameter of the bond is 8 cm while its thickness is 1.3 mm. The Young's modulus and bulk modulus of the adhesive were measured to be 3.45 MPa, 575 MPa, respectively.

We first compute the diameter-to-thickness ratio, D/t, as

$$\frac{D}{t} = \frac{80\text{mm}}{1.3\text{mm}} \approx 60 .$$

The Poisson's ratio, ν, is computed from Eq. (3.19) as

$$\nu = \frac{1}{2} - \frac{E}{6B} = \frac{1}{2} - \frac{\left(3.45\ \mu\text{N/mm}^2\right)}{6\left(575\ \mu\text{N/mm}^2\right)} = 0.499 ,$$

and the shear modulus, G, is computed from Eq. (3.20) as

$$G = \frac{E}{2(1+\nu)} = \frac{\left(3.45\ \mu\text{N/mm}^2\right)}{2\left[1+(0.499)\right]} = 1.15\ \mu\text{N/mm}^2.$$

The maximum modulus, M, given by Eq. (3.18) is calculated as

$$M = \frac{(1-\nu)E}{(1+\nu)(1-2\nu)} = \frac{(1-0.499)3.45\times10^6\,\mu\text{N/mm}^2}{(1+0.499)\left[1-2(0.499)\right]} = 5.765\times10^8\,\mu\text{N/mm}^2$$

With Table 3.5 for $\nu = 0.499$ and D/t=60, k_{33} can be interpolated as

$$k_{33} = \frac{60-50}{100-50}(0.7574-0.5580)+0.5580 = 0.5979.$$

Similarly, k_{31} can be interpolated as

$$k_{31} = \frac{60-50}{100-50}(0.9981-0.9953)+0.9953 = 0.9959.$$

Since the adhesive bonds are shaped by the spherical form of the back surface of the optic, the material coordinate system is chosen to be a spherical system centered at the pads' center of curvature. Since the through-the-thickness direction is in the radial direction of this spherical coordinate system, the terms in Eq. (3.22) must be reorganized to the following form:

$$\begin{Bmatrix} \sigma_{11} \\ \sigma_{22} \\ \sigma_{33} \\ \tau_{12} \\ \tau_{23} \\ \tau_{31} \end{Bmatrix} = \begin{bmatrix} k_{33}M & \dfrac{k_{31}k_{33}\nu M}{(1-\nu)} & \dfrac{k_{31}k_{33}\nu M}{(1-\nu)} & 0 & 0 & 0 \\ \dfrac{k_{31}k_{33}\nu M}{(1-\nu)} & M & \dfrac{\nu M}{(1-\nu)} & 0 & 0 & 0 \\ \dfrac{k_{31}k_{33}\nu M}{(1-\nu)} & \dfrac{\nu M}{(1-\nu)} & M & 0 & 0 & 0 \\ 0 & 0 & 0 & G & 0 & 0 \\ 0 & 0 & 0 & 0 & G & 0 \\ 0 & 0 & 0 & 0 & 0 & G \end{bmatrix} \begin{Bmatrix} \varepsilon_{11} \\ \varepsilon_{22} \\ \varepsilon_{33} \\ \gamma_{12} \\ \gamma_{23} \\ \gamma_{31} \end{Bmatrix}.$$

Substitution of values gives a material matrix as follows:

$$\begin{bmatrix} 3.447\times10^8 & 3.419\times10^8 & 3.419\times10^8 & 0 & 0 & 0 \\ 3.419\times10^8 & 5.765\times10^8 & 5.742\times10^8 & 0 & 0 & 0 \\ 3.419\times10^8 & 5.742\times10^8 & 5.765\times10^8 & 0 & 0 & 0 \\ 0 & 0 & 0 & 1.15\times10^6 & 0 & 0 \\ 0 & 0 & 0 & 0 & 1.15\times10^6 & 0 \\ 0 & 0 & 0 & 0 & 0 & 1.15\times10^6 \end{bmatrix} \mu\text{N/mm}^2.$$

3.2.3.3 Effective properties for ring bonds

Figure 3.23 shows an example of a ring bond. The form of the effective properties for this type of bond is shown in Eq. (3.24). While the "1" direction is in the radial direction through the thickness of the bond, the "2" and "3" directions are in the hoop and axial directions, respectively.

$$\begin{Bmatrix} \sigma_{11} \\ \sigma_{22} \\ \sigma_{33} \\ \tau_{12} \\ \tau_{23} \\ \tau_{31} \end{Bmatrix} = \begin{bmatrix} k_{11}M & \dfrac{k_{12}k_{11}\nu M}{(1-\nu)} & \dfrac{k_{13}k_{11}\nu M}{(1-\nu)} & 0 & 0 & 0 \\ \dfrac{k_{12}k_{11}\nu M}{(1-\nu)} & k_{11}M & \dfrac{k_{13}k_{11}\nu M}{(1-\nu)} & 0 & 0 & 0 \\ \dfrac{k_{13}k_{11}\nu M}{(1-\nu)} & \dfrac{k_{13}k_{11}\nu M}{(1-\nu)} & k_{33}M & 0 & 0 & 0 \\ 0 & 0 & 0 & G & 0 & 0 \\ 0 & 0 & 0 & 0 & G & 0 \\ 0 & 0 & 0 & 0 & 0 & G \end{bmatrix} \begin{Bmatrix} \varepsilon_{11} \\ \varepsilon_{22} \\ \varepsilon_{33} \\ \gamma_{12} \\ \gamma_{23} \\ \gamma_{31} \end{Bmatrix}. \quad (3.24)$$

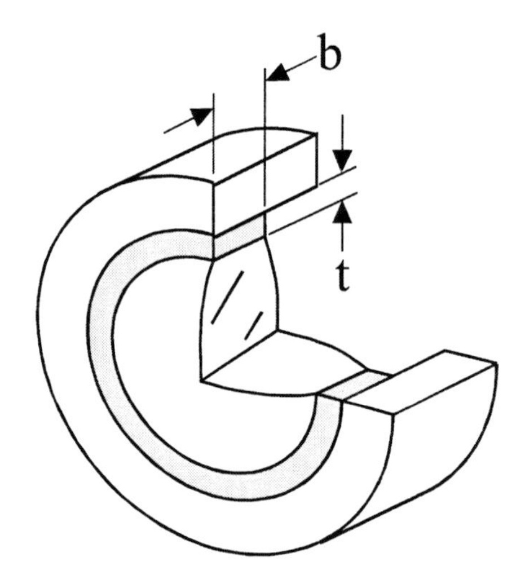

Figure 3.23 Example of a ring bond design with thickness, *t*, and width, *b*.

The correction factors k_{22}, k_{12}, k_{13}, and k_{33} are tabulated in Table 3.6 for various b/t ratios and Poisson's ratios. The effective properties for ring bonds are insensitive to the ratio of the radius of the ring bond to its thickness (R/t) for ratios above 10. Therefore, these effective properties may also be used for a very long straight bond in which the "1" direction is through the thickness of the bond, the "2" direction is in the long dimension, and the "3" direction is along the width of the bond.

When employing effective properties of bonds in simplified solid models, the user must be aware of their limitation. The effective properties to be used with coarse-bond models are simply intended to match the overall stiffness of the bond. Since regions of the bond closer to the free faces will display more compliance in the actual hardware, there may be a significant variation in stiffness throughout the bond. Such variation in stiffness may be important to the behavior of the hardware and must be included in the model in such cases. Use of the effective properties, however, prevents representation of such a distribution of stiffness. Therefore, in ring-bond designs like that shown in Figure 3.24(a) a detailed model of the bond must be employed. However, designs like those shown in Figure 3.24(b) have been found to be accurately represented by coarse solid models and effective properties.

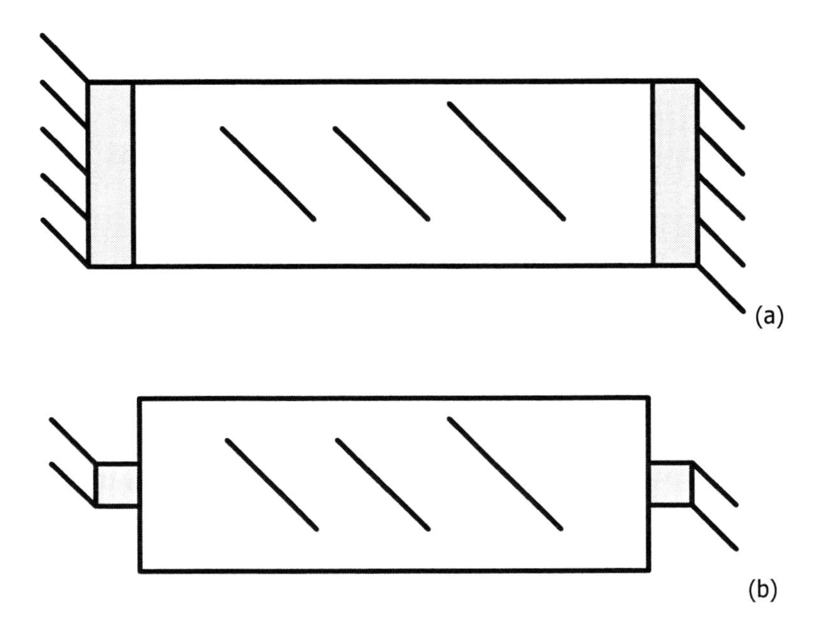

Figure 3.24 Cross-section plots of two example ring bond designs: (a) full width bond and (b) partial width bond.

Table 3.6 Correction factors for ring bonds with various combinations of *b/t* ratio and Poisson's ratio.

	$\nu = 0.45$				$\nu = 0.49$				$\nu = 0.499$				$\nu = 0.4999$			
B/T RATIO	K_{11}	K_{12}	K_{13}	K_{33}	K_{11}	K_{12}	K_{13}	K_{33}	K_{11}	K_{12}	K_{13}	K_{33}	K_{11}	K_{12}	K_{13}	K_{33}
1	0.4036	0.6717	0.2704	0.1433	0.1018	0.6399	0.2652	0.0355	0.0108	0.6328	0.2641	0.0037	0.0011	0.6321	0.2640	0.0004
2	0.5101	0.7866	0.5257	0.3750	0.1484	0.7657	0.5219	0.1099	0.0165	0.7611	0.5213	0.0123	0.0017	0.7607	0.5213	0.0012
5	0.7521	0.9267	0.8372	0.8518	0.3665	0.9295	0.8560	0.4523	0.0554	0.9316	0.8630	0.0703	0.0058	0.9319	0.8637	0.0074
10	0.8760	0.9685	0.9301	0.9907	0.6295	0.9760	0.9510	0.8248	0.1691	0.9803	0.9605	0.2386	0.0205	0.9809	0.9617	0.0292
20	0.9383	0.9854	0.9675	0.9994	0.8126	0.9906	0.9808	0.9871	0.4169	0.9944	0.9888	0.5922	0.0742	0.9950	0.9900	0.1086
50	0.9756	0.9944	0.9876	0.9952	0.9252	0.9967	0.9933	1.0002	0.7460	0.9986	0.9973	0.9616	0.3155	0.9991	0.9983	0.4580
100	0.9883	0.9974	0.9941	0.9904	0.9628	0.9984	0.9968	0.9955	0.8730	0.9994	0.9988	0.9999	0.6005	0.9997	0.9995	0.8299
200	0.9954	0.9990	0.9977	0.9941	0.9820	0.9993	0.9985	0.9897	0.9365	0.9997	0.9995	0.9994	0.7977	0.9999	0.9998	0.9883
500	0.9991	0.9998	0.9996	0.9987	0.9949	0.9998	0.9996	0.9951	0.9749	0.9999	0.9998	0.9899	0.9192	1.0000	0.9999	1.0001
1000	0.9997	0.9999	0.9999	0.9996	0.9985	0.9999	0.9999	0.9985	0.9891	1.0000	0.9999	0.9915	0.9597	1.0000	1.0000	0.9948

3.3 Displacement Models of Flexures and Mounts

3.3.1 Classification of mounts

Mounts can be classified in terms of how they are linked to their surroundings. All mounts can be grouped into one of the following types.

Unstable refers to mounting schemes that fail to react at least one rigid-body motion of the mounted structure, as illustrated in Fig. 3.25(a). In a static finite element solution the stiffness matrix representing an unstable structure is singular. Therefore, attempts to obtain a static solution of displacements to any applied loading will result in division by zero.

Kinematic refers to a mounting scheme that reacts all rigid body motions of the mounted structure with no redundancy as illustrated in Fig. 3.25(b). The reactions to the structure at the kinematic points of contact may be determined without regard to the knowledge of the stiffness of the structure, or of the surroundings to which the structure is mounted. A kinematically mounted structure is isolated from elastic deformations of its surroundings, although it may undergo rigid-body motion as its surroundings deform and move.

Redundant refers to a mounting scheme that elastically couples a structure to its surroundings as illustrated in Fig. 3.25(c). Such a structure will elastically deform when its surroundings are elastically deformed, and such deformations are dependent on the stiffness of the structure and the surroundings.

Pseudo-kinematic is a term referring to a special case of redundant mounting also called flexured. Pseudo-kinematic mounts are weakly redundant mounting schemes that attempt to approximate a kinematic mounting scheme as illustrated in Fig. 3.25(d). The redundancies are minimized by designing flexures that exhibit relatively large stiffness only in directions where kinematic constraints would be applied.

3.3.2 Modeling of kinematic mounts

Although perfectly kinematic mounts are not achievable in practice, it is often useful to idealize a mounting interface as kinematic for an analysis. Such an approach allows an analyst to simplify the model used for an early design trade study or to bound the displacement prediction by eliminating the redundant mounting effects. Kinematic mounts are modeled with either constraints or rigid

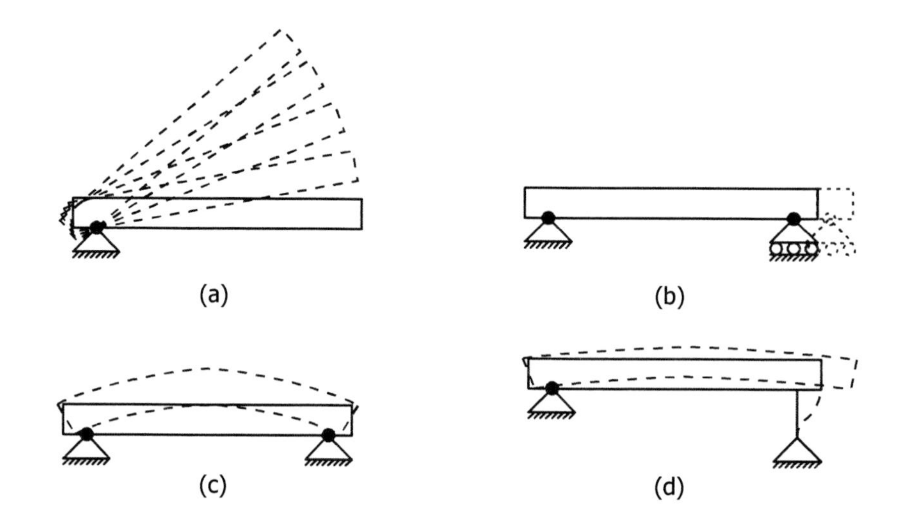

Figure 3.25 Classification of mounts: (a) unconstrained, (b) perfectly kinematic, (c) redundant, and (d) pseudo-kinematic or flexured.

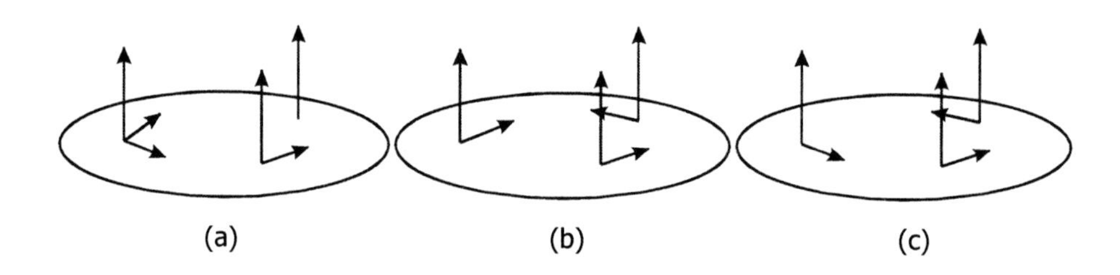

Figure 3.26 Example kinematic mounting schemes: (a) cone, groove, and flat, (b) three grooves, (c) three grooves in a modified configuration.

elements. Constraints are used if the surroundings are not included in the model, while rigid elements are employed when interfacing the models of two components.

Both methods of modeling kinematic mounts must be defined so that the directions they constrain or link are correctly represented. Figure 3.26 shows three mounting schemes using the same set of three mounting point locations but using different constrained directions. Each of these kinematic mounts will behave differently; therefore, it is important to correctly represent the intended scheme. Each finite element code uses its own method of defining the direction in which constraints will act; thus, it is important that the analyst follow the method properly.

In addition to the defined directions of the kinematic constraints, the defined locations of the nodes to which the constraints are applied are equally important. Figure 3.27 shows an optic mounted with kinematic mounts located at its midplane in one case and at its backplane in a second case. As the illustration

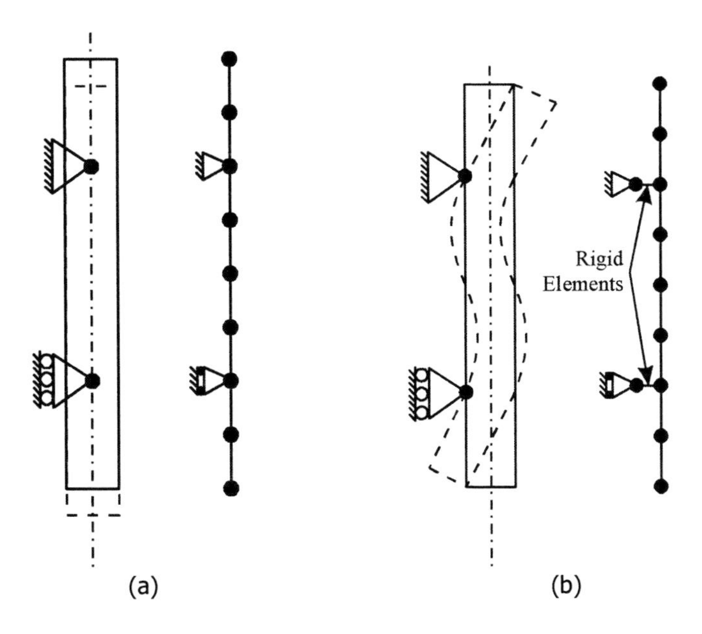

Figure 3.27 Effect of kinematic constraint location on a laterally loaded mirror: (a) kinematic constraints located on the neutral plane, and (b) kinematic constraints located off the neutral plane.

shows, the locations of the kinematic-mount points affect the resulting deformed shape; therefore, they must be properly represented. If the construction of the model is such that finite element nodes are not defined where the kinematic mount points are located, then rigid elements may be used to link the mount points to the model, allowing the constraints to be applied in the proper location as shown in Fig. 3.27(b).

3.3.3 Modeling of flexure mounts

If the goal of an analysis warrants such detail, flexure mounts may be modeled to include their redundant stiffness characteristics. Inclusion of such stiffnesses will show, for example, the transmitted moments of a flexure-mounted optic whose metering structure undergoes elastic deformation.

3.3.3.1 Modeling of beam flexures

Beam flexures exhibit significant stiffness only along their axes. The bending and transverse shear stiffnesses are relatively small. In situations where beam flexures are used in pairs, as shown in Fig. 3.28, the location of the strut intersection points (SIP) are very important to the behavior of the mounted optic. In addition, the orientation of the bipod pair can be an important factor. The importance of properly modeling the location of the SIP and the orientation of the bipod pair is analogous to the importance of properly modeling the locations and reaction directions of kinematic mounts discussed above.

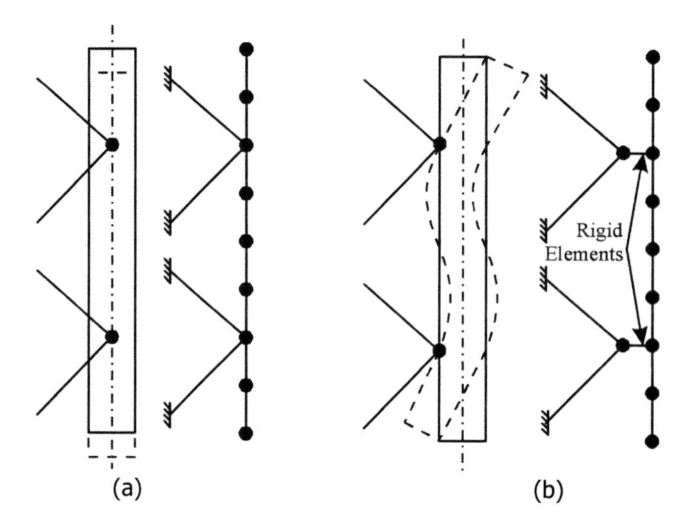

Figure 3.28 Effect of bipod flexure strut-intersection point (SIP) location on a laterally loaded mirror: (a) SIP located on the neutral plane, and (b) SIP located off the neutral plane.

In order to correctly represent the bending and transverse shear stiffnesses of beam flexures it is important to choose the proper active flexure length to use in the model. The length of the active flexures in the hardware is often not well defined due to the fillets at each end. Therefore, effective lengths must be chosen to represent the active flexures and their fillets. Figure 3.29 shows an example beam-flexure design and the corresponding finite element beam model. Proper representation of the active flexure length in most designs can be achieved by including half of each of the fillet lengths in the flexure portions of the mesh. The nominal beam properties of the active flexure are then assigned to this effective length.

When performing design-trade studies on beam-flexure bipods, it is helpful to organize the flexure models so they can be easily modified in a text editor. If the finite element nodes of a beam-flexure bipod are defined in a cylindrical coordinate system as shown in Fig. 3.30, then the analyst can change the

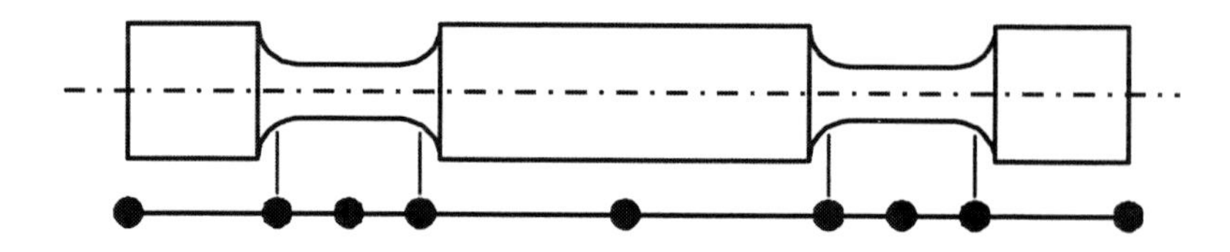

Figure 3.29 Example beam-flexure design and corresponding finite element mesh. The modeled active flexure lengths include half of each fillet length.

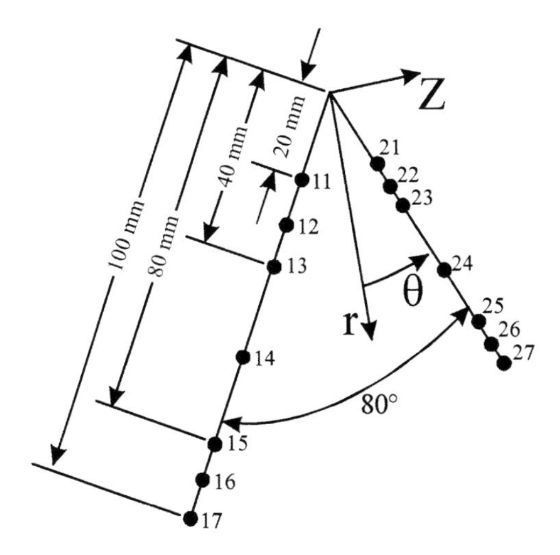

Figure 3.30 Finite element model of beam flexure bipod showing the defining coordinate system to allow easy changes to the model. See Table 3.7 for a corresponding list of coordinate locations.

dimensions along the length of the flexure by editing the radial coordinate locations. Furthermore, the spread angle of the bipod flexures can be altered simply by changing the azimuthal coordinate locations. Also, the slant angle of the bipod-pair plane is defined solely by the orientation of the *Z* axis of the cylindrical coordinate system. Table 3.7 shows the coordinate locations of the numbered nodes in Fig. 3.30 as an example. Changes to this model description are more easily made in a text editor with column select-and-replace features than in a graphical preprocessor.

Table 3.7 Cylindrical coordinates of flexure bipod model nodes shown in Fig. 3.30.

NODE ID	R (MM)	θ (°)	Z (MM)
11	20.0	–40.0	0.0
12	30.0	–40.0	0.0
13	40.0	–40.0	0.0
14	60.0	–40.0	0.0
15	80.0	–40.0	0.0
16	90.0	–40.0	0.0
17	100.0	–40.0	0.0
21	20.0	40.0	0.0
22	30.0	40.0	0.0
23	40.0	40.0	0.0
23	60.0	40.0	0.0
25	80.0	40.0	0.0
26	90.0	40.0	0.0
27	100.0	40.0	0.0

3.3.3.2 Example: modeling of bipod flexures

Finite element models of the beam flexures used to mount a mirror are to be included in an analysis that predicts the optical-surface deformation due to enforced motion of the flexure ends. Such motion may be associated with thermoelastic expansion of the metering structure to which the flexures are bonded, or due to locked-in strain during assembly of the flexures to the metering structure. The flexures, whose dimensions are shown in Fig. 3.31, are to be fabricated of titanium. The mirror is identical to that used in Example 3.1.4.4. The enforced displacements should be 0.01 in. and applied to each flexure base in the direction normal to the plane defined by each bipod. The flexures are modeled with solid elements in one analysis and with beam elements in a second analysis to provide a comparison.

The finite element meshes of the two model types are shown in Fig. 3.32. While the bar-element model includes 26 nodes and 24 elements per bipod, the solid element model contains 118,530 nodes and 115, 200 elements per bipod. A comparison of the results from each analysis is given in Table 3.8.

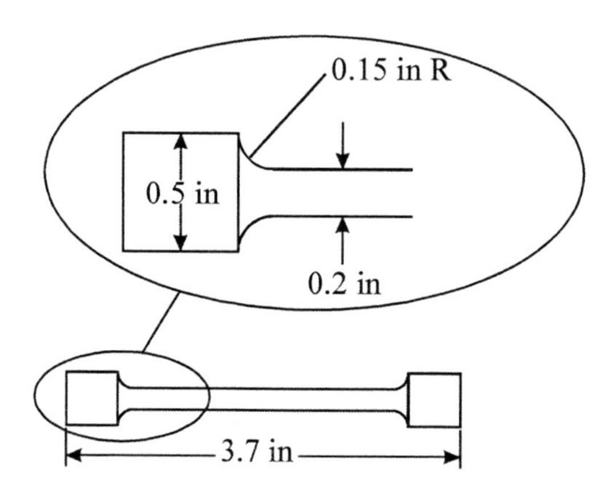

Figure 3.31: Dimensions of example beam-flexure design.

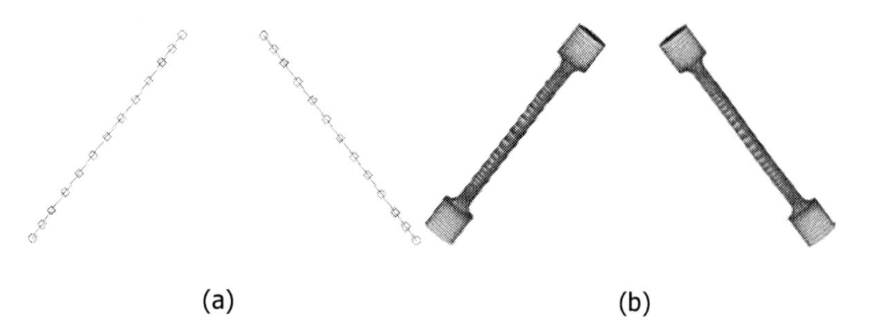

Figure 3.32 Finite element models of example beam flexures: (a) beam model and (b) solid model.

Table 3.8 Comparison of surface deformation results computed with different flexure models.

	BAR-ELEMENT FLEXURES			SOLID-ELEMENT FLEXURES		
	MAG. (NM)	RESIDUAL RMS (NM)	RESIDUAL P-V (NM)	MAG. (NM)	RESIDUAL RMS (NM)	RESIDUAL P-V (NM)
Input		1065.8	609.2		1056.6	598.8
Bias	1052.9	149.5	609.2	1043.9	146.9	598.8
Power	239.4	59.5	324.9	235.3	58.5	319.3
Pri Trefoil	151.3	25.9	149.8	148.8	25.4	147.2
Pri Spherical	−14.8	24.7	149.8	−14.5	24.3	147.2
Sec Trefoil	32.7	23.1	146.6	32.2	22.7	144.1
Sec Spherical	−29.2	20.4	123.9	−28.7	20.1	121.8
Pri Hexafoil	34.5	18.2	113.3	33.9	17.9	111.4
Ter Trefoil	58.7	10.0	71.8	57.7	9.8	70.6
Ter Spherical	5.2	9.8	70.9	5.1	9.7	69.7
Sec Hexafoil	11.7	9.4	66.9	11.5	9.2	65.8

Notice that the results correlate very well, illustrating that the bar-element model is equally as capable in describing the stiffness of the bipod flexures as the solid-element model. This excellent correlation results from the fact that the assumptions made in using the bar element models to represent the bipod flexures were very valid assumptions for this problem. Furthermore, the bar-element model provides a more effective tool for predicting certain results such as the moments transmitted into the mirror. However, the reader should not dismiss the use of high-fidelity models when they are necessary. The key point to be conveyed by this example is that by understanding the capabilities of each modeling method and by anticipating the mechanical behavior of the system to be analyzed, the most cost effective approach can be chosen to accomplish the goals of an analysis.

3.3.3.3 Modeling of blade flexures

Blade flexures such as those shown in Fig. 3.33 are often not pseudo-kinematic. Redundant loads in the form of moments about the displayed X and Z-axes can be significant. They are best suited for situations where such redundant loads are not likely to be generated. Proper modeling of these flexures, however, should include such redundant stiffnesses.

Bar elements can be used to model blade flexures in some cases but should usually be limited to design trade study analyses instead of final verification analyses. Bar elements give the analyst the advantage of easily verifying the absence of redundant loads by requesting the forces in the bar elements representing the flexure. However, while a bar-element mesh may provide a first-order representation of the flexure stiffness, the bar stresses should not be considered accurate. Plate- and shell-element meshes of blade models will more correctly represent both the stiffness and stresses for final verification analyses.

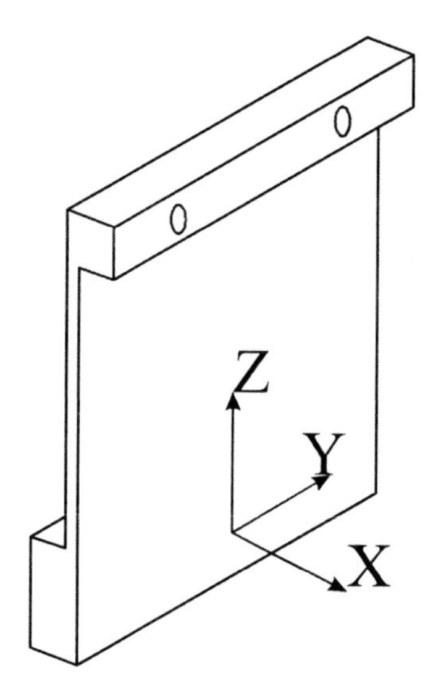

Figure 3.33 Example blade flexure.

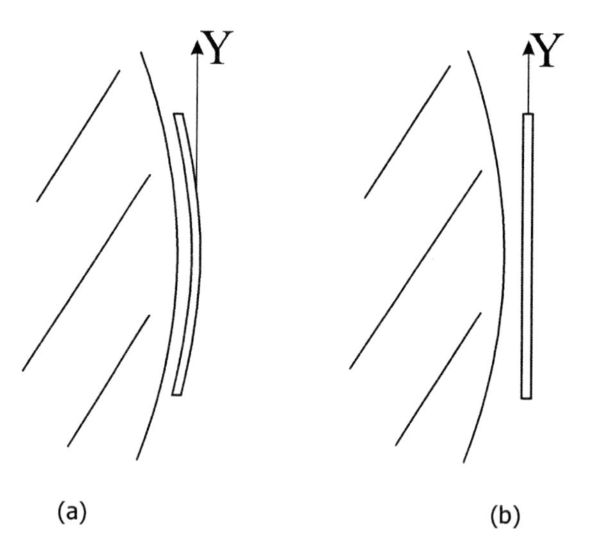

Figure 3.34 Two variations of blade flexures that have different bending stiffnesses about the *Y* axis: (a) curved blade and (b) flat blade.

If bar-element meshes are to be used to represent blade flexures, the analyst must be careful to calculate the properties correctly. If the flexures are sections of a cylindical shell as shown in Fig. 3.34(a), the section properties of the hardware illustrated in Fig. 3.34(b) may not correctly represent the bending stiffness about the *Y* axis. The curved geometry of the flexure cross-section can often add significant bending stiffness to the flexure.

3.3.4 Modeling of test supports

The purpose of performing a test-support deformation analysis is often to assess the surface-error contribution due to fabricating an optic to a desired prescription while in a test support that does not adequately represent the optics in-use support. This error contribution, however, is as much a function of the optic in its in-use configuration as it is a function of the optic in its test support. Figure 3.35 illustrates that an optic is fabricated to its prescription while being supported by an air bag. It is then subsequently mounted in an inclined configuration on its mounts. The error contribution of interest is the difference between the deformed optical surfaces of these two states. Since a linear finite element analysis assumes that the model begins in a stress-free and strain-free state, the deformation analysis of the optic in each state is the deformation change relative to a perfectly figured optic floating in a zero-gravity (0-g) environment. To obtain the change in surface figure between two deformed states, a node-by-node difference in the finite element displacement results must be generated before surface characterization is performed. Some finite element codes allow users to accomplish such a difference operation within the finite element analysis. However, a simple program or spreadsheet application can be used to difference the results of two analysis cases.

A test-support deformation analysis may be important, if surface figuring methods such as ion figuring or small tool polishing are employed. The inverse of the shift in surface figure from the test-support configuration to the in-use

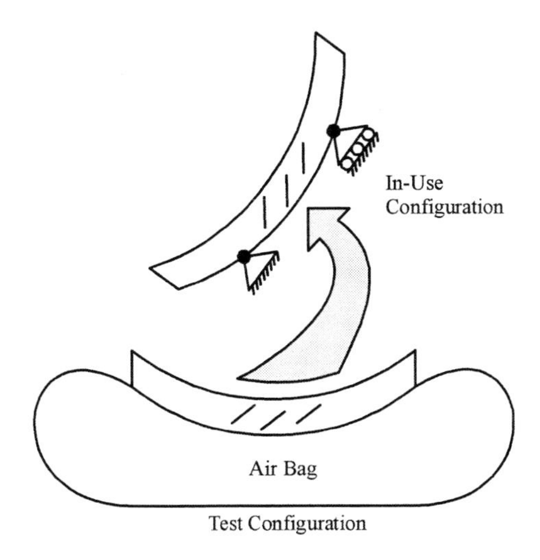

Figure 3.35 The optic is figured to the environment in which it is tested, and it can display different figures in its operational environments.

configuration can be fabricated into the surface of the optic, thereby lessening the effects caused by testing the optic in an environment different from the in-use environment. This process is accomplished by generating an analytically computed array representing the deformation change caused by the test support and adding this array to the interferogram results of each test measurement performed during fabrication of the optic. As each figuring pass is performed, the optical figure will converge to the desired prescription minus the anticipated deformation change. The surface figure error contribution associated with going from the test state to the in-use state would then be the error with which the analytical prediction was made.

In addition, various optical testing procedures require limits on the deviation of the optical surface from its intended shape. Such requirements may impose restrictions on how the test support should be designed to adequately support the optic or optical system so that accurate test results can be obtained. Therefore, analysis prediction of how optical systems deform in their test supports can be very important.

3.3.4.1 Modeling of air bags

Air bags are commonly used to simulate a 0-g environment during an optical test. Methods of modeling air bags stem from the fact that the pressure inside the air bag is either assumed constant or is a function of the hydraulic head, h, as illustrated in Fig. 3.36. Therefore, for an axisymmetric optic supported by an air bag, the air bag can be represented by a uniform pressure applied normal to the supported face of the optic. This method, however, assumes that test engineers have inflated the air bag such that tangency is achieved at all points around the edge of the optic as shown in Fig. 3.37(a). If the air bag is underinflated or overinflated, then tangency will not be achieved as shown in Figs. 3.37(b) and (c). If the optic has a center hole, then a properly sized weight can be placed in the center hole so that the air bag is forced to become tangent at both the inner and outer edges. Therefore, the analyst is encouraged to communicate with test engineers who are responsible for the design and use of the test support hardware in order to understand what effects may need to be modeled.

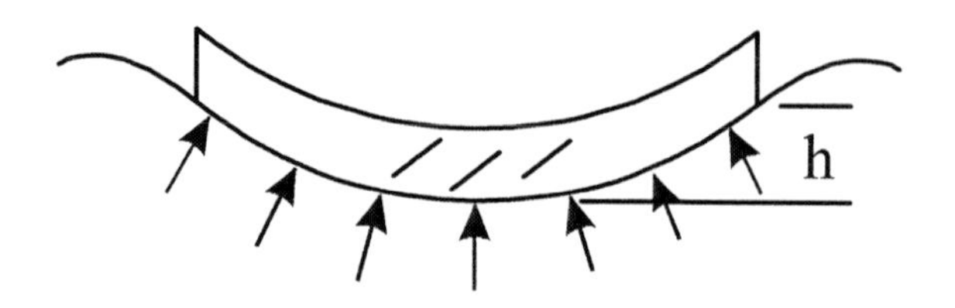

Figure 3.36 An air bag can be modeled as a representative pressure that reacts the weight of the optic.

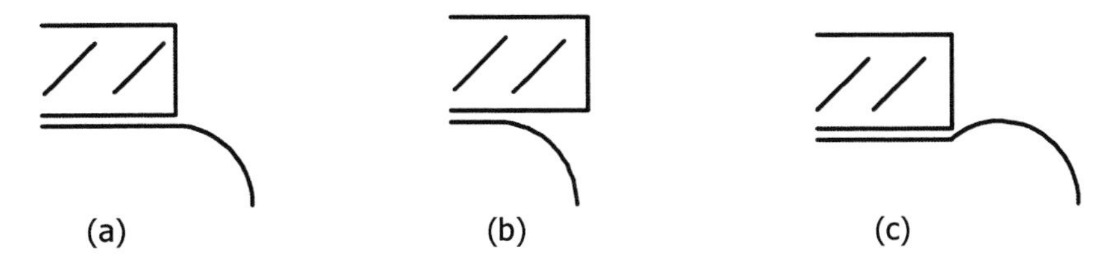

Figure 3.37 Three edge conditions of an air bag support: (a) tangency, (b) overinflated, and (c) underinflated.

If tangency can be assumed everywhere, the method for computing the proper pressure to apply is as follows:

<1> With a finite element analysis, compute the model weight, W, and the net load in the direction of gravity, F_p, generated by a unit pressure load applied normal to the supported face of the optic.

<2> Compute the pressure, p, which will identically balance the weight, W, by the following equation:

$$p = \frac{W}{F_p}. \tag{3.25}$$

<3> Apply kinematic constraints, the pressure, p, and the gravity load in a static-finite element analysis. Request recovery of the reactions and verify that they are zero.

The rigid-body motions predicted by this analysis are arbitrary, but the elastic shape of the optical surface after its rigid-body motions are removed will be a reliable prediction.

If the optic is not axisymmetric, as shown in the highly exaggerated illustration in Fig. 3.38, then tangency cannot be achieved at all points around the edge of the optic. This lack of tangency can be modeled as a varying line load applied to the optic edge wherever tangency is lacking. The methods for computing the pressure and line load are as follows:

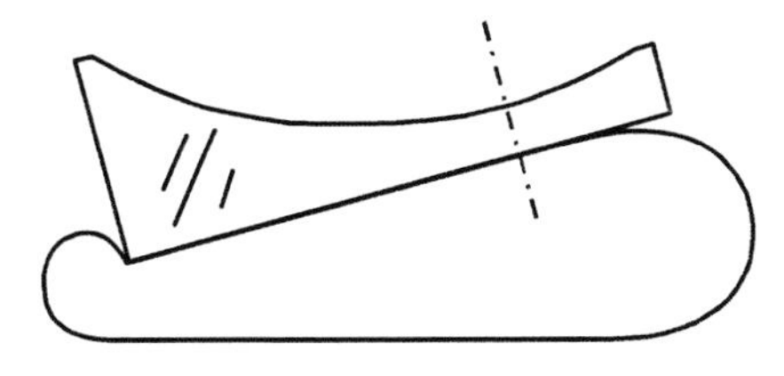

Figure 3.38 Exaggerated illustration of a nonaxisymmetric mirror supported by an air bag.

≺1≻ With a finite element analysis compute both the net vertical load and net moment generated by each of the following: the model weight, a unit pressure load applied normal to the supported face of the optic, and a line load $u(\theta)$ given by

$$u(\theta) = a\cos(\theta) + b\,, \tag{3.26}$$

where the constants a and b are chosen such that $u(0)$ is unity, and its values of zero are located at points where tangency is achieved. Figure 3.39 shows an illustration of what this line load may look like. Table 3.9 defines the values that are calculated in this step.

≺2≻ The pressure, p, and line-load peak, ω_0, which balance the weight can be found by the following:

$$p = \frac{-M_\omega W - F_\omega M_W}{-F_p M_\omega + F_\omega M_p}, \tag{3.27}$$

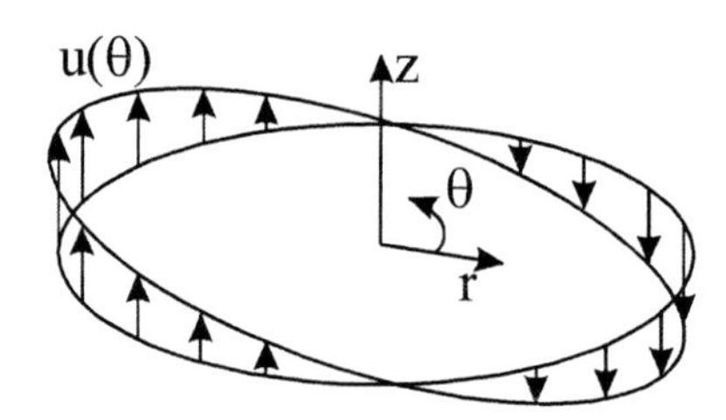

Figure 3.39 Varying line load representing the lack of tangency around the periphery of a nonaxisymmetric optic supported by an air bag.

Table 3.9 Net-load values computed for nonaxisymmetric optic supported by an air bag.

	WEIGHT	UNIT PRESSURE	UNIT LINE LOAD
Net Vertical Load	W	F_p	F_ω
Net Moment	M_W	M_p	M_ω

$$\omega_0 = \frac{M_p W + F_p M_W}{-F_p M_\omega + F_\omega M_p}. \tag{3.28}$$

The line-load function $\omega(\theta)$ then becomes

$$\omega(\theta) = \omega_0 u(\theta) = \omega_0 \left[a\cos(\theta) + b \right]. \tag{3.29}$$

<3> Apply kinematic constraints, the pressure, p, the line load $\omega(\theta)$, and the gravity load in a static finite element analysis. Request recovery of the reactions and verify that they are zero. Reactions that are nonzero indicate an error in the application of the loads or an error in how they were computed.

An assumption inherent to the calculations shown above is that any areas of "lift off" as illustrated in Fig. 3.37(b) are small compared to the supported surface of the optic.

3.3.4.2 Example: test support deformation analysis of a nonaxisymmetric optic

An off-axis mirror whose cross section is shown in Fig. 3.40 is to be tested on an air bag during fabrication but is to be mounted for operation in a 1-g environment on three back-surface points for operation. The surface error generated by transferring the optic from the air bag to its mounted in-use configuration is desired so that it may be included in the wavefront-error budget for the system. The mirror is fabricated of a glass (ULE) whose material properties are given in Table 3.10.

A solid-element model of the mirror is shown in Fig. 3.41. Two analysis cases will be performed on this model to predict the surface error change. The first case involves finding the deformation of the optic on the air bag relative to a 0-g environment. We first must calulate the proper unit edge load, $u(\theta)$. The test engineers have specified that tangency will be achieved at two points shown in Fig. 3.42. Therefore, from Eq. (3.26) we write

$$u(0\ \text{deg}) = a\cos(0\ \text{deg}) + b = 1,$$

and

$$u(37\ \text{deg}) = a\cos(37\ \text{deg}) + b = 0.$$

The constants a and b are obtained by solving a set of two simultaneous equations:

$$\begin{bmatrix} \cos(0^\circ) & 1.0 \\ \cos(37^\circ) & 1.0 \end{bmatrix} \begin{Bmatrix} a \\ b \end{Bmatrix} = \begin{Bmatrix} 1.0 \\ 0.0 \end{Bmatrix},$$

$$\begin{bmatrix} 1.0 & 1.0 \\ 0.799 & 1.0 \end{bmatrix} \begin{Bmatrix} a \\ b \end{Bmatrix} = \begin{Bmatrix} 1.0 \\ 0.0 \end{Bmatrix},$$

$$\begin{Bmatrix} a \\ b \end{Bmatrix} = \begin{Bmatrix} 4.975 \\ -3.975 \end{Bmatrix}.$$

Table 3.10 Material properties of ULE.

PROPERTY NAME	PROPERTY VALUE
Young's Modulus	6.757×10^{10} μN/mm^2
Poisson's Ratio	0.17
Mass Density	2.187 g/cm^3

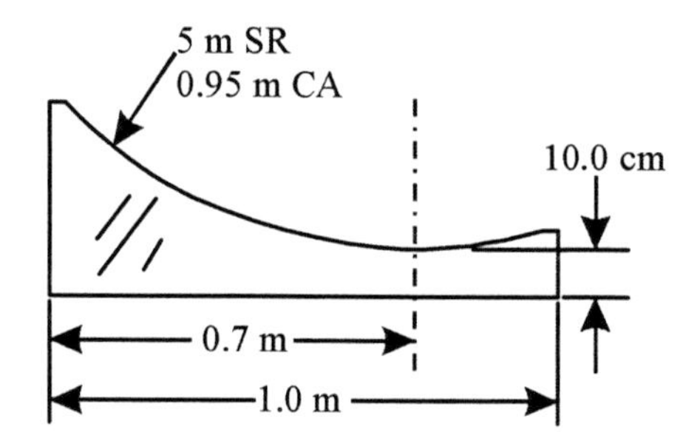

Figure 3.40 Dimensions of an off-axis mirror to be tested on an air bag during fabrication.

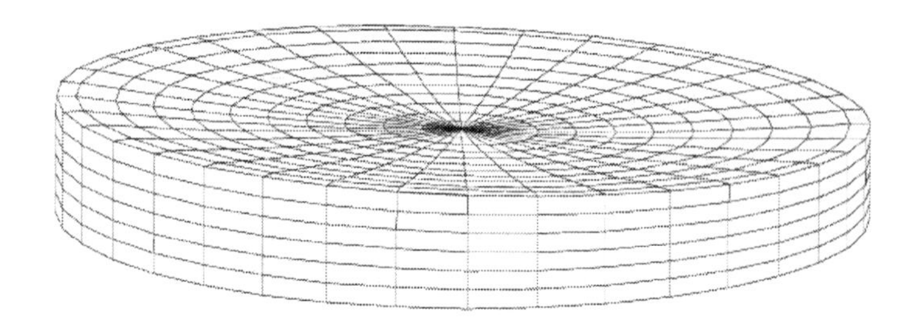

Figure 3.41 Finite element model of the off-axis mirror shown in Fig. 3.40.

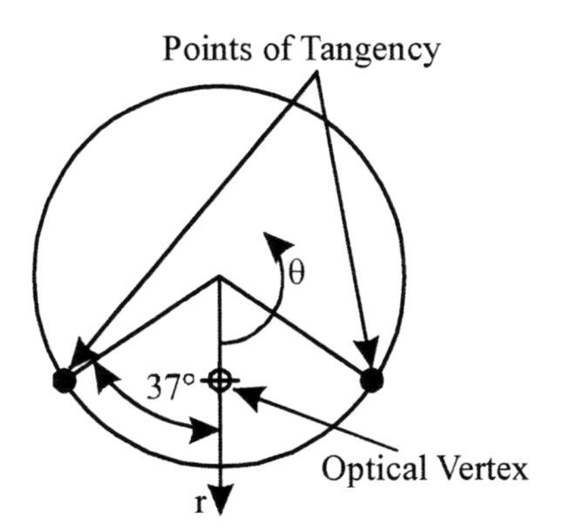

Figure 3.42 Locations of tangency between the air bag and the off-axis mirror.

Therefore, the unit line load becomes

$$u(\theta) = 4.975\cos(\theta) - 3.975.$$

The net-vertical loads and moments about the optical-surface vertex are found by an initial finite element analysis with kinematic constraints for the application of gravity, a constant back pressure, and the unit line load found above. These net loads are shown in Table 3.11.

Table 3.11 Net loads for weight, unit pressure, and unit line load.

	WEIGHT	UNIT PRESSURE	UNIT LINE LOAD
Net vertical load	**1.950×10^9 N**	**7.814×10^5 N**	**-1.247×10^4 N**
Net moment	**-4.163×10^{10} N/mm**	**0.0 N/mm**	**-3.883×10^6 N/mm**

With the values in Table 3.11, and Eqs. (3.27), and (3.28), we can compute the pressure, p, and the line-load peak, ω_0, to exactly balance the weight of the mirror.

$$p = \frac{-M_\omega W - F_\omega M_W}{-F_p M_\omega + F_\omega M_p} = \frac{-\left(-3.883\times10^6\right)\left(1.950\times10^9\right) - \left(-1.247\times10^4\right)\left(-4.163\times10^{10}\right)}{-\left(7.814\times10^5\right)\left(-3.883\times10^6\right) + \left(-1.247\times10^4\right)\left(0.0\right)}$$

$$= 2{,}324.4\frac{N}{\text{mm}^2},$$

$$\omega_0 = \frac{M_p W + F_p M_W}{-F_p M_\omega + F_\omega M_p} = \frac{(0.0)(1.950\times10^9)+(7.814\times10^5)(-4.163\times10^{10})}{-(7.814\times10^5)(-3.883\times10^6)+(-1.247\times10^4)(0.0)}$$

$$= -10,721.1\frac{N}{\text{mm}}.$$

The line load applied to the outer edge is redefined with Eq. (3.29) to become

$$\omega(\theta) = \omega_0 u(\theta) = -10,721.1[4.975\cos(\theta) - 3.975]$$
$$= 53,337.5\cos(\theta) - 42,616.4\frac{N}{\text{mm}}.$$

The displacements due to the gravity load, constant pressure, p, and line load, $\omega(\theta)$, are found with the kinematic-boundary conditions applied in a second finite element analysis. The air bag loads balance the vertical load and moment from the weight to within 162 μN and 1936 μN/mm, respectively. These imbalances are very small, indicating that we have computed the effective air bag loads correctly.

The next step involves finding the deformation change between a zero gravity environment and the in-use mounted configuration. The mounts are idealized as kinematic mounts at the three mount locations and displacements due to a gravity load applied along the optical axis are requested in a third analysis.

A node-by-node difference is performed to subtract the displacements associated with the air bag case from the displacements of the in-use mounted case. The rigid-body motions of the surface are then extracted from the nodal displacement differences, and the residual rms surface error is found.

The surface error results from the air bag case, the in-use case, and the difference are summarized in Table 3.12.

Table 3.12 Surface error results.

	SURFACE RMS (NM)	SURFACE P-V (NM)
Air-bag loads	**11 nm**	**47 nm**
In-use loads	**114 nm**	**499 nm**
Difference	**116 nm**	**511 nm**

The system engineer then doubles the 116-nm-rms prediction for inclusion in the wavefront-error budget.

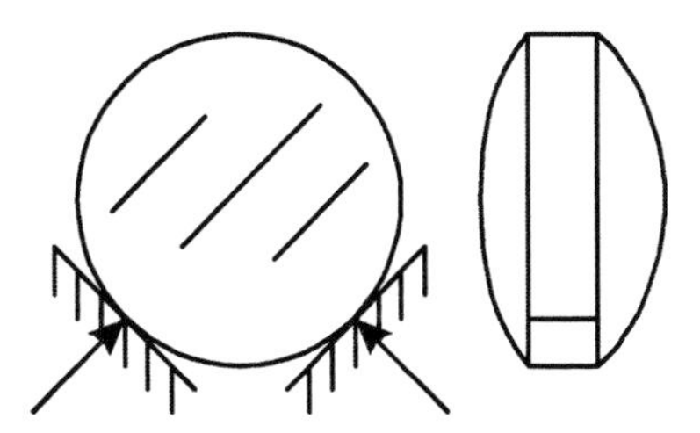

Figure 3.43 V-block supports are modeled by correctly oriented line constraints.

3.3.4.3 Modeling of V-block test supports

The modeling of a V-block test support such as that shown in Fig. 3.43 can be performed by applying constraints to the optic along a line contact. If a frictionless surface is to be assumed, then the constraints must be oriented so that they only constrain displacements normal to the optic. For circular optics, the simplest way of assuring this is to define the constraints in the radial direction of a cylindrical-coordinate system that is located on the axis of the optic. The analyst should be careful to construct the model of the optic so that a line of nodes is located at the line contact representing the V-block. In addition to the line constraints representing the V-block contacts, the analyst must also include enough additional constraints to precisely remove the component rigid-body motions along and about the optical axis. The reactions associated with these constraints, as predicted by the finite element analysis, should be verified to be zero.

3.3.4.4 Modeling of sling and roller-chain test supports

A sling or roller-chain support such as that shown in Fig. 3.44 can be modeled by a pressure load given by

$$p_0 = \frac{W}{Dt}, \tag{3.30}$$

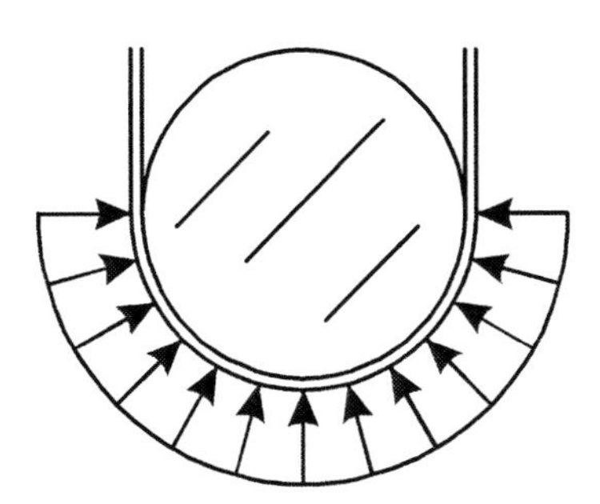

Figure 3.44 Sling supports are modeled by a constant pressure that reacts the optic's weight.

where D is the diameter of the optic, and t is the width of the contact area between the sling or roller chain, and the optic edge. As in other test-support modeling methods, enough constraints must be applied to precisely remove any singular rigid-body motions. The reactions associated with these constraints as predicted by the finite element analysis should be verified to be zero.

3.4 Displacement Analysis Methods

This section deals with deformation analysis methods for applications that may not be as obvious as simple gravity-deformation analysis, for example.

3.4.1 Analysis of surface effects

The causes of surface effects that induce deformations of optics include thermoelastic mismatching between an optic and its coating, coating-cure shrinkage, coating-moisture absorption, and the Twyman effect. The Twyman effect is the deformation of an optic after release of the surficial stress-state generated by the polishing process. Such a release can be caused by abraiding its polished surface. All of these surface effects can cause optical-surface deformation and stresses in the surface coatings that can lead to cracking.

All of the surface effects mentioned above can be simulated with a thermoelastic analysis. In the cases of coating shrinkage, moisture absorption, and the Twyman effect, effective-thermoelastic strains, α_c*, are computed from Table 3.13. These effective-thermoelastic strains can be applied to the model by using α_c* as the CTE for the coating and a unit temperature change. Details for each modeling method are given below.

3.4.1.1 Composite-plate model

Some finite element codes have composite property features with which the user may specify the material and thickness of each layer of a composite-layer stack. This feature can be used to predict surface-deformation effects in plate models representing an optic and its coating. The optic and its coating are modeled with one layer of plate elements, which is given a composite-property description as illustrated in Fig. 3.45. The property values assigned to each layer of the composite description are shown in Table 3.14.

One advantage of the composite plate model is that the stresses in the surface layer and the interlaminar shear stresses may be recovered. In cases where the surface layer represents a surface coating, this may be useful information if cracking of the coating is suspected.

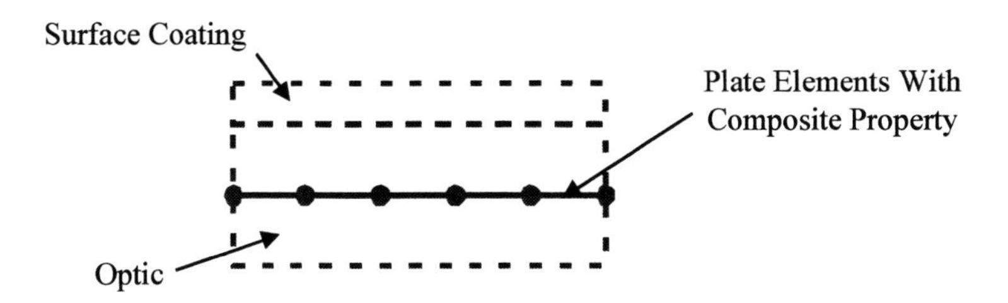

Figure 3.45 Composite-plate model for surface-effect deformation prediction.

Table 3.13 Effective $\alpha\Delta T$ for surface effects.

COATING EFFECT	EFFECTIVE $\alpha\Delta T$ OF COATING (α_c*)
Moisture Absorption Growth	$CME\,\Delta M$
Cure Shrinkage	$-\dfrac{\sigma_c(1-\nu_c)}{E_c}$
Twyman Effect[9]	$\dfrac{4C}{3t_c}$

CME is the coefficient of moisture expansion of the coating, and ΔM is the moisture change value in units consistent with CME.
σ_c is the stress in the coating deposited on a rigid substrate.
ν_c and E_c are the Poisson's ratio and Young's modulus, respectively. Assume $\nu_c = 0$ for worst case if ν_c is not known.
C is the Twyman constant.
t_c is an arbitray small thickness which must also be used in the finite element model if applicable.

Table 3.14 Composite-layer property values for surface-effect analysis.

	OPTIC LAYER				SURFACE LAYER				
COATING EFFECT	E	ν	CTE	T	E	ν	CTE	T	TEMPERATURE LOAD
Thermoelastic	E_o	ν_o	α_o	t_o	E_c	ν_c	α_c	t_c	ΔT
Moisture absorp.	E_o	ν_o	0	t_o	E_c	ν_c	α_c^*	t_c	Unity
Cure shrinkage	E_o	ν_o	0	t_o	E_c	ν_c	α_c^*	t_c	Unity
Twyman effect	E_o	ν_o	0	t_o	E_o	ν_o	α_c^*	Arbitrary	Unity

3.4.1.2 Homogeneous-plate model

Since many finite element codes do not feature composite-property descriptions, results for surface deformation effects in plate models can be obtained with a homogeneous-plate model and effective thermoelastic loads. Such a model is illustrated in Fig. 3.46. The properties of the homogeneous plate model are simply unmodified values one would use for any other type of analysis. The effective thermoelastic loads, however, include a bulk temperature shift and a thermal gradient to include the surface effect. Equations [3.31(a)] and [3.31(b)] are used to compute the loads for thermoelastic analyses. Equation [3.31(a)] gives the bulk temperature shift, $\overline{\Delta T}$, and temperature gradient, T', which should be applied to the homogeneous model. The equivalent coated and uncoated surface temperatures, ΔT_1 and ΔT_2, for codes that require such a format are given in Equation [3.31(b)].

$$\Delta T^* = \frac{E_c \alpha_c \Delta T}{E_o \alpha_o}, \quad \overline{\Delta T} = \frac{\Delta T t_o + \Delta T^* t_c}{t_o}, \quad T' = \frac{6 \Delta T^* t_c}{t_o^2}, \qquad [3.31(a)]$$

$$\Delta T_1 = \overline{\Delta T} - \frac{3 \Delta T^* t_c}{t_o}, \quad \Delta T_2 = \overline{\Delta T} + \frac{3 \Delta T^* t_c}{t_o}, \qquad [3.31(b)]$$

Symbols used in Equations [3.31(a)] and [3.31(b)] are defined as follows:

$\overline{\Delta T}$ = effective bulk temperature shift;
T' = effective thermal gradient;
ΔT_1 = effective temperature of uncoated surface;
ΔT_2 = effective temperature of coated surface;
E_c = Young's modulus of coating;
E_o = Young's modulus of optic;
α_c = CTE of coating;
α_o = CTE of optic;
ΔT = temperature change;
t_c = thickness of coating; and
t_o = thickness of optic.

These equations assume that the plate elements are defined such that a positive value of T' will generate a thermoelastic load with the highest temperature on the coated side. It is also assumed that t_c is much smaller than t_o.

For analagous analyses such as those listed in Table 3.13, the following equations are used:

$$\Delta T^* = \frac{E_c \alpha_c^*}{E_o \alpha_o}, \quad \overline{\Delta T} = \frac{\Delta T^* t_c}{t_o}, \quad T' = \frac{6 \overline{\Delta T}}{t_o}, \qquad [3.32(a)]$$

and

$$\Delta T_1 = \overline{\Delta T} - 3 \Delta T^*, \quad \Delta T_2 = \overline{\Delta T} + 3 \Delta T^*, \qquad [3.32(b)]$$

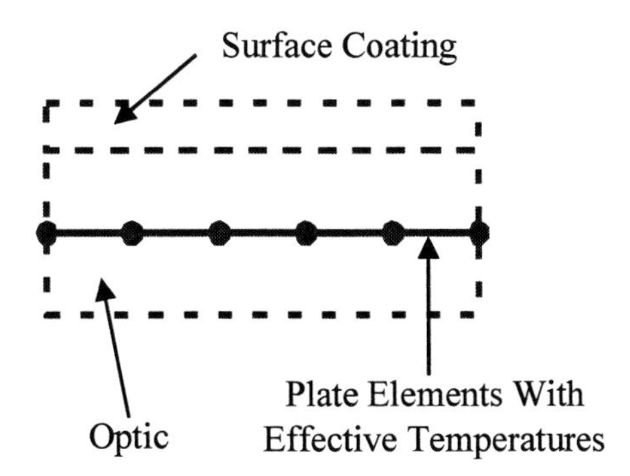

Figure 3.46 Homogeneous-plate model for surface-effect deformation prediction.

where α_c* is one of the effective $\alpha\Delta T$ values found in Table 3.11, and α_o is an arbitrary CTE, which must also be used in the optic's finite element model material description. Notice that for simulation of the Twyman effect, t_c is an arbitrary value.

This model lacks the material properties of the coating layer. Therefore, stresses in the coating layer are not correctly predicted by the homogeneous-plate model.

3.4.1.3 Three-dimensional model

For optics that require 3D-solid models, such as thick lenses, surface effects can be included by a mesh of membrane elements on the optic's surface as shown in Fig. 3.47. The definition of properties for the solid and surface meshes are similar to the methods used for the composite plate description discussed in Sec. 3.4.1.1. Table 3.15 gives the correct property values for the solid-element mesh of the optic and the membrane-element mesh of the surface where α_c* is found in Table 3.13.

Table 3.15 Three-dimensional-model property values for surface effect analysis.

	OPTIC MESH			SURFACE MESH				
COATING EFFECT	E	ν	CTE	E	ν	CTE	T	TEMPERATURE LOAD
Thermoelastic	E_o	ν_o	α_o	E_c	ν_c	α_c	t_c	ΔT
Moisture absorp.	E_o	ν_o	0	E_c	ν_c	α_c^*	t_c	Unity
Cure shrinkage	E_o	ν_o	0	E_c	ν_c	α_c^*	t_c	Unity
Twyman effect	E_o	ν_o	0	E_o	ν_o	α_c^*	Arbitrary	Unity

Notice that the stresses predicted by the membrane elements can be recovered to compute the stress in the coating.

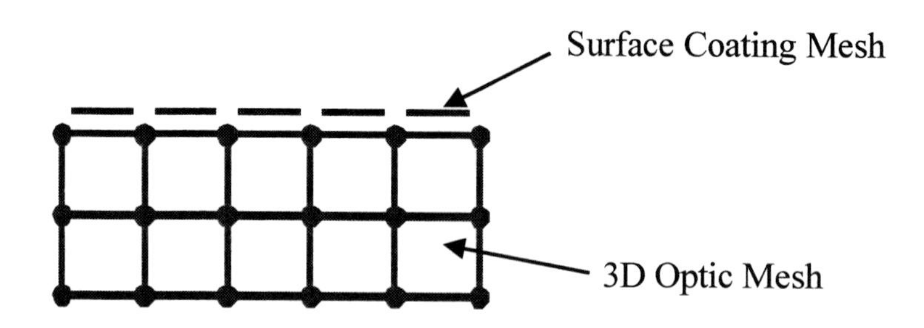

Figure 3.47 Three-dimensional model for surface-effect deformation prediction.

3.4.1.4 Example: coating-cure shrinkage

A reflective coating is deposited onto the optical surface of a solid 1.0-in.-thick flat mirror fabricated of a glass whose Young's modulus is 13.2 msi and Poisson's ratio is 0.272. The Young's modulus of the coating is 1500 psi, and its thickness is 0.0001 inch. The coating stress is 2100 psi. The change in surface figure is desired due to the cure shrinkage of the coating layer. The results from the three methods of analysis (composite plate, homogeneous plate, and 3D solid) are to be used to compute the surface error for comparison.

From Table 3.13 we can compute the effective thermoelastic load, $\alpha\Delta T = a_c^*$, to apply to the coating:

$$\alpha_c^* = -\frac{\sigma_c}{E_c} = -\frac{(2100\text{psi})}{(1500\text{psi})} = -1.4$$

We apply this effective-thermoelastic load by setting the CTE of the coating to −1.4, setting the CTE of the optic to 0.0, and applying a unit increase in temperature to the model in a thermoelastic analysis.

3.4.1.4(a) Composite-plate model

The composite-plate-model property description defines a 1.0-in. layer with a Young's modulus of 13.2 msi, and a CTE of 0.0 laminated to a second layer 0.0001-in. thick with a Young's modulus of 1500 psi, and a CTE of −1.4. The MSC/NASTRAN[8] PCOMP property and material descriptions that accomplish this are listed below.

```
PCOMP  10001                                70.0
       10001  1.0      0.0  NO      12001   0.0001   0.0   YES
MAT1   10001  13.2+6        0.272           0.0      70.0
MAT1   12001  1.5+3         0.272          -1.4      70.0
```

The optic mesh is supported by kinematic constraints and a unit temperature drop is applied to the model as an isothermal thermoelastic load.

3.4.1.4(b) Homogeneous-plate model

The homogeneous-plate model simply consists of a plate mesh whose property is a constant 1.0-in. thickness, and whose material properties are those of the glass. The element normals are defined so that a positive thermal gradient is consistent with a higher temperature on the coated surface as compared to the uncoated surface. The effective temperatures, which include the effect of the coating, are computed from

$$\Delta T^* = \frac{E_c \alpha_c^*}{E_o \alpha_o} = \frac{(1500\text{psi})(-1.4)}{(13.2\times 10^6\text{psi})(3.78\times 10^{-6}/{}^\circ\text{F})} = -42.1\,{}^\circ\text{F},$$

$$\overline{\Delta T} = \frac{\Delta T^* t_c}{t_o} = \frac{(-42.1\,{}^\circ\text{F})(0.0001\,\text{in.})}{(1.0\,\text{in.})} = -0.0042\,{}^\circ\text{F},$$

and

$$T' = \frac{6\overline{\Delta T}}{t_o} = \frac{6(-0.0042\,{}^\circ\text{F})}{1.0\ \text{in.}} = -0.0252\,{}^\circ F/\text{in.}$$

The optic mesh is supported by kinematic constraints and the computed temperature loads are applied in a thermoelastic analysis.

3.4.1.4(c) Three-dimensional model

A solid mesh of the 1.0-in.-thick optic is created and membrane elements are added to represent the coated surface. The material properties of the solid elements are those of the glass. The thickness and material properties of the membrane element are the same as those of the coating. However, the CTE of the solid elements is set to 0.0, while the CTE of the membrane elements is set to −1.4. The optic mesh is supported by kinematic constraints, and a unit temperature increase is applied in a thermoelastic analysis.

The analysis results of the three methods are summarized in Table 3.16.

Table 3.16 Result summary of coating-cure analyses.

MODEL TYPE	TOTAL RMS	TOTAL P-V	P-V POWER
Composite plate	**70.3 nm**	**121.2 nm**	**121.2 nm**
Homogeneous plate	**70.3 nm**	**121.3 nm**	**121.3 nm**
Three-dimensional	**69.7 nm**	**121.1 nm**	**120.8 nm**

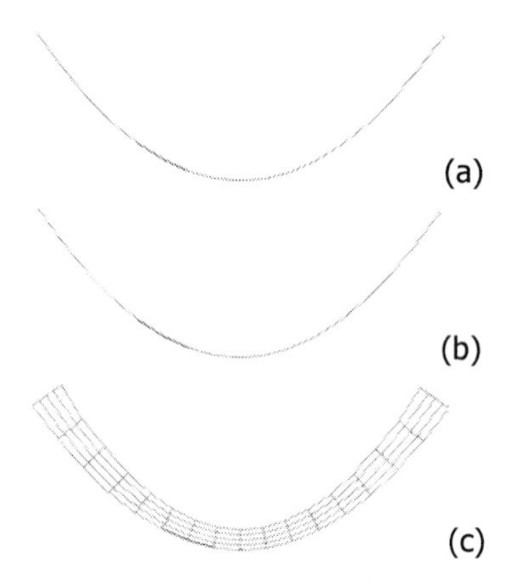

Figure 3.48 Exaggerated deformed shapes of optic after coating-cure shrinkage: (a) composite-plate model, (b) homogeneous-plate model, and (c) 3D model.

The results in Table 3.16 show an excellent correlation between all three model types. The 3D model shows some compliance associated with through the thickness deformation near the edges of the optic. Figure 3.48 shows the exaggerated deformed shapes for the diametrical cross sections of each model type.

Both the composite-plate and 3D models predict a coating stress of 2885 psi. The homogeneous-plate model is unable to predict the coating stress.

3.4.2 Analysis of assembly processes

In many applications, optical performance can be affected by deformations that are locked into an optical system during its assembly. Therefore, it is of interest in many situations to be able to predict how much deformation will result from a particular process of assembling an optical system. Figure 3.49 shows an illustration of a simple assembly process in which a mirror is bonded to its mounts in a 1-g environment and subsequently placed in a 0-g operational environment. The process of bonding the mirror to its flexures while being supported by the assembly fixture locks in elastic strain, which remains in the unloaded final state. Figure 3.50 shows an illustration of the surface rms error at each step in the process. Notice that the change in deformation between states can be found through a linear finite element analysis of the system with changes in externally applied loads or internal connections. The analysis of the whole process, however, can be performed with a piecewise nonlinear analysis.[10] Therefore, a nonlinear analysis capability is required in which separate load steps can be defined to vary loads, and to alter model connections. Furthermore, each load step must be applied to the deformed model of each previous load step.

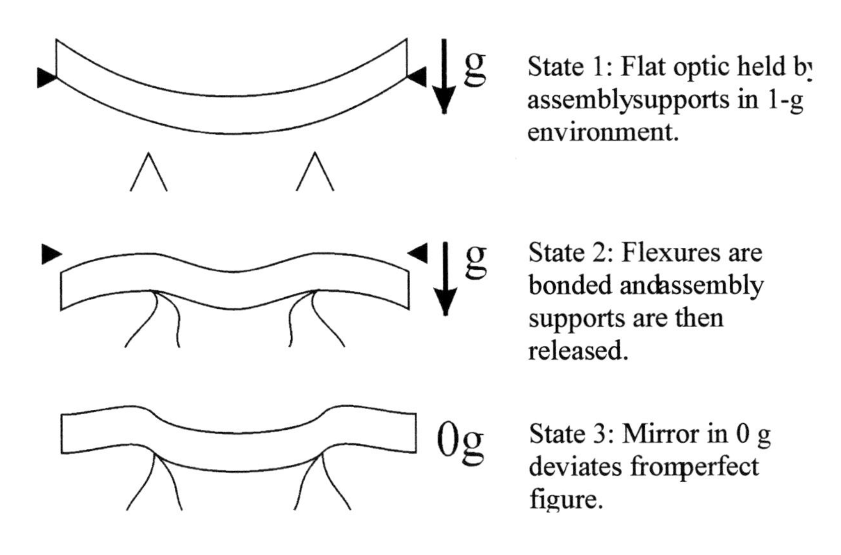

Figure 3.49 Assembly process for bonding an optic to flexure mounts.

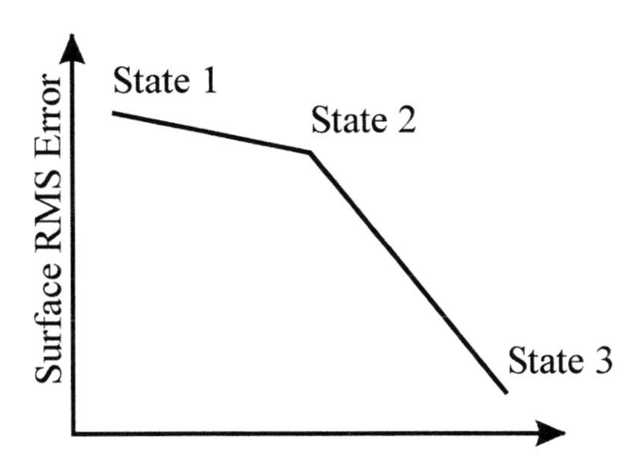

Figure 3.50 Surface rms error relative to a strain-free state through the assembly process illustrated in Fig. 3.49.

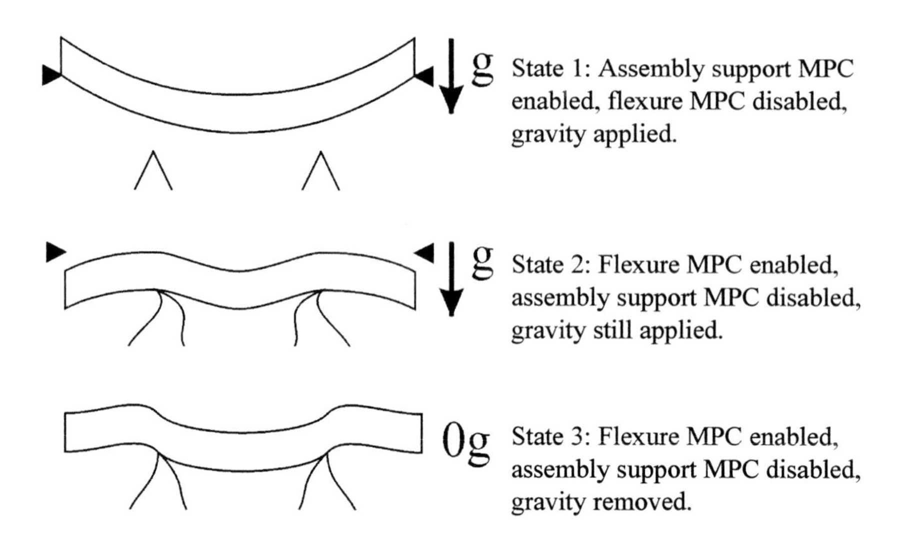

Figure 3.51 Analysis flow for simulation of assembly process for bonding an optic to flexure mounts.

Figure 3.51 illustrates how this analysis would be executed in a finite element analysis. Three load steps would be defined in a nonlinear analysis. Multi-point constraints (MPC) are used to model the connections to the assembly support and flexures. These connections must have the capability of being active or inactive in each load step. The first load step activates the MPC between the optic and the assembly support and applies the gravity load. The MPC between the optic and the flexures, however, is left inactive. The second load step activates the MPC between the optic and the flexures, and deactivates the MPC between the optic and the assembly support. The gravity load is left on. This simulates the process of bonding the optic to its mounts and transfers the weight of the optic to the flexures. The third load step keeps the MPC between the optic and the flexures active while removing the gravity load. This load step simulates the process of transferring the assembled system to a 0-g environment while having been bonded in a 1-g environment.

While the analysis demonstrated in Fig. 3.51 is relatively simple, much more complex processes can be modeled by adding more steps and connection interfaces to the assembly analysis.

3.4.2.1 Example: assembly analysis of mirror mounting

The assembly process illustrated in Fig. 3.49 is to be performed on the mirror used in Example 3.1.4.4. The partial listing of the MSC/NASTRAN model below shows the analysis control section (lines between "CEND" and "BEGIN BULK") and some key entries in the bulk data section (lines below "BEGIN BULK"). Comments are included in italic to clarify the purpose of important entries.

```
SOL 106 NONLINEAR STATICS ANALYSIS
CEND
TITLE = 3D plate MIRROR
SUBTITLE = ASSEMBLY ANALYSIS
ECHO = NONE
NLPARM = 1        NONLINEAR PARAMETER REQUEST
DISP(PUNCH,PLOT) = ALL NODAL DISPLACEMENT REQUEST
SUBCASE 1         BEFORE ASSEMBLY SUBCASE - NO MIRROR TO FLEXURE MPC
 LOAD = 1         GRAVITY LOAD
 SPC = 1          FLEXURE BASE AND ASSEMBLY SUPPORT CONSTRAINED
SUBCASE 2         AFTER ASSEMBLY SUBCASE
 LOAD = 1         GRAVITY LOAD
 SPC = 2          ONLY FLEXURE BASE CONSTRAINED
 MPC = 2          CONNECTION BETWEEN FLEXURES AND MIRROR TURNED ON
SUBCASE 3         GRAVITY REMOVED - SHOWS LOCKED IN STRAIN EFFECTS
 SPC = 2          ONLY FLEXURE BASE CONSTRAINED
 MPC = 2          CONNECTION BETWEEN FLEXURES AND MIRROR KEPT ON
BEGIN BULK
NLPARM  1         NONLINEAR ANALYSIS PARAMETERS - USE ALL DEFAULTS
$ GRAVITY LOAD
GRAV    1    0     386.4  0.0    0.0  -1.0
$ FLEXURE BASE AND ASSEMBLY SUPPORT CONSTRAINTS FOR 1ST SUBCASE
SPC1    1    123456 300113 300213 300313 300413 300513 300613
SPC1    1    23    105618 105582 105600
$ FLEXURE BASE CONSTRAINTS FOR 2ND AND 3RD SUBCASES - NO ASSEMBLY
SUPPORT
SPC1    2    123456 300113 300213 300313 300413 300513 300613
MPC     2   300011 1    -1.0  300001 1     1.0
$ FLEXURE TO MIRROR CONNECTIONS FOR 2ND AND 3RD SUBCASES
MPC     2   300012 1    -1.0  300002 1     1.0
MPC     2   300013 1    -1.0  300003 1     1.0
.
.
MPC     2   300011 6    -1.0  300001 6     1.0
MPC     2   300012 6    -1.0  300002 6     1.0
MPC     2   300013 6    -1.0  300003 6     1.0
```

Figure 3.52 shows the deformed shape after gravity has been removed from the assembled system. Although the residual elastic deformation of the mounted mirror is negligible, 1200 nm of piston is the result of the initial deformation of the flexures before bonding.

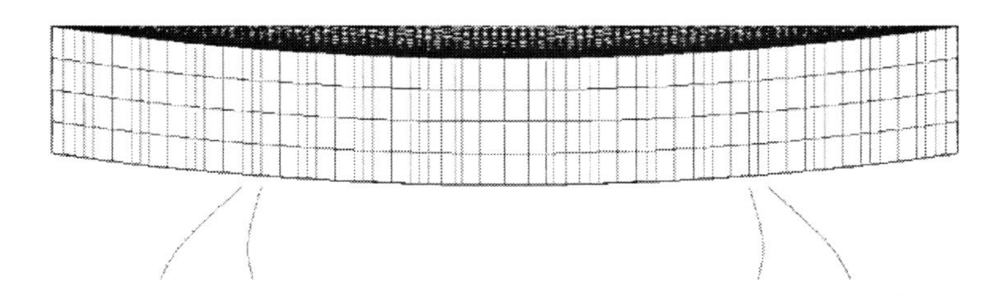

Figure 3.52 Highly exaggerated deformed shape of lightweight mirror in 0-g environment, which was mounted in 1-g environment.

References

1. Young, Warren C., *Roark's Formulas for Stress and Strain, Sixth Ed.*, McGraw-Hill, Inc., New York (1989).
2. Jones, Robert M., *Mechanics of Composite Materials*, McGraw-Hill, Inc., New York (1975).
3. ULE is a trademark of Corning, Inc.
4. Lindley, P.B., *Engineering Design with Natural Rubber*, Natural Rubber Technical Bulletin, 3rd Edition, published by the National Rubber Producers Research Association (1970).
5. Tsai, Hsiang-Chuan and Chung-Chi Lee, "Compressive Stiffness of Elastic Layers Bonded between Rigid Plates," *Int. J. Solids Structures*, **35**(23), pp.3053–3064 (1998).
6. Genberg, V. L., "Structural Analysis of Optics," SPIE Short Course (1986).
7. Michels, G. J., V. L. Genberg, K. B. Doyle, "Finite element modeling of nearly incompressible bonds,"
8. MSC/NASTRAN is a product of the MacNeal-Schwendler Corporation, Los Angeles, CA.
9. Rupp, W. J., "Twyman effect for ULE," *Proceedings of Optical Fabrication and Testing Workshop*, pp. 25–30, OSA, Washington, D.C. (1987).
10. Stone, M. J., V. L. Genberg, "Nonlinear superelement analysis to model assembly process," *Proceedings of MSC World Users Conference* (1993).

≺Chapter 4≻ Integration of Optomechanical Analyses

This chapter discusses the integration of finite element derived response quantities such as displacements and stress into the model. These techniques allow mechanical design trades to be performed as a function of optical performance. Also discussed are methods to compute line-of-sight jitter and to evaluate the effects of mechanical obscuration on image quality.

4.1 Optical Surface Positional Errors

Mechanical loads acting on an optical system can significantly degrade optical performance by changing the position of the optical surfaces. Positional or rigid-body errors include translations and rotations of a surface in six degrees-of-freedom. Translation of the optic along the optical axis is called despace, changes in lateral position are called decenter, and tip and tilt refer to rotations about the lateral axes. These positional errors are shown in Fig. 4.1. For nonrotationally symmetric optics, rotation about the optical axis must also be considered. Common optical errors resulting from shifts in the position of an optical surface include focus, astigmatism, and coma. In addition, a boresight or a line-of-sight error may result as illustrated in Fig. 4.2.

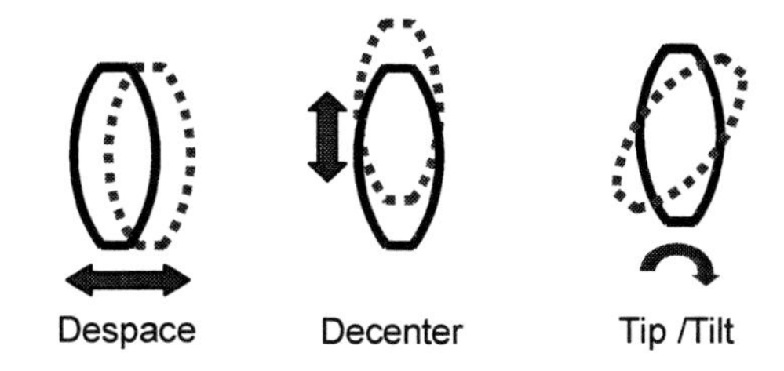

Figure 4.1 Rigid-body optical element motions.

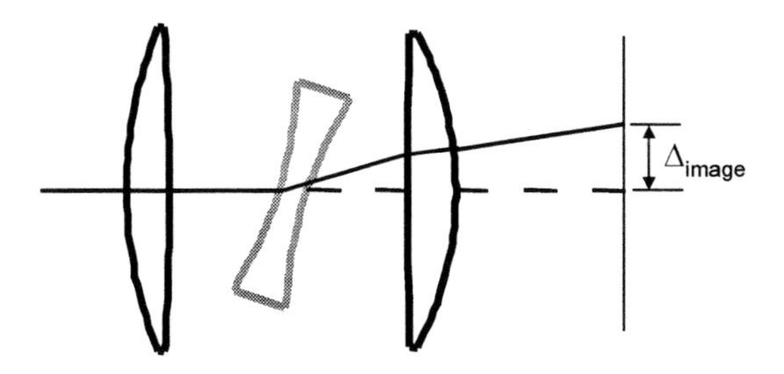

Figure 4.2 Line-of-sight error is the lateral displacement of the on-axis image point.

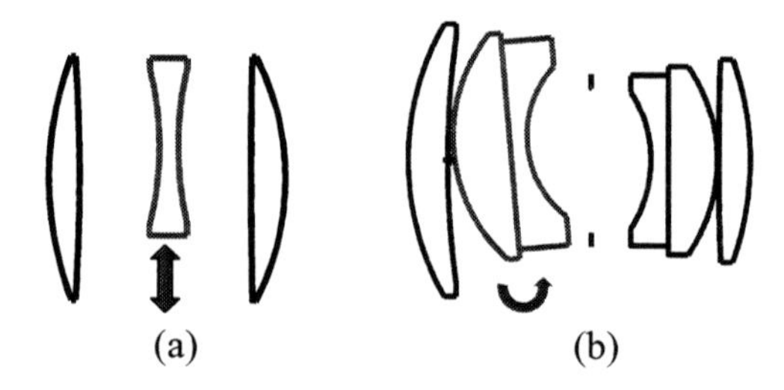

Figure 4.3 (a) Decenter of a single lens element, and (b) tilt of a cemented doublet.

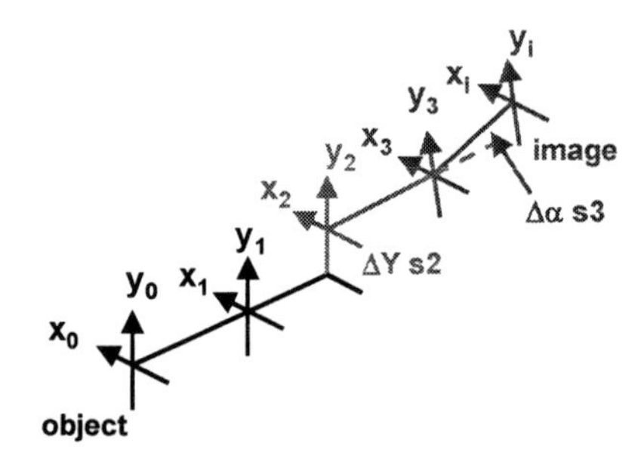

Figure 4.4 In the optical model, decenters and tilts are applied to the local coordinate system defining the surface.

Rigid-body errors may be accounted for in the optical model by perturbing individual or groups of surfaces as shown in Fig. 4.3. This requires relating the mechanical coordinate system in which the rigid-body motions are computed and the coordinate system used to define the optical surface in the optical model. Typically, the optical surface is defined at the location of the vertex. For on-axis optics, where the vertex is at the geometric center of the optic, computing the motion of the vertex is straightforward. For off-axis optics where the vertex is physically not on the optical substrate, coordinate transformations are required.

Applying rigid-body errors to optical surfaces is done by tilting or decentering the local coordinate system that defines the surface as illustrated in Fig. 4.4. Nominally, this results in cumulative errors since each local coordinate system is defined relative to the local coordinate system of the proceeding surface. A common method to uncouple the perturbations is to specify a decenter and return, which as the name implies, returns the local coordinate system of the surface following the tilted and decentered surface to the original coordinate system. Repeating this command for each of the surfaces allows rigid-body errors to be defined independently. Other methods to uncouple the rigid-body errors include use of global coordinates in CODE V and coordinate breaks in ZEMAX.[1,2] In general, applying rotations to an optical surface in the optical model is order dependent. However, for small rotations such as those typically computed by a linear finite element analysis, the rotations are order independent.

4.2 Optical Surface Shape Changes

Mechanical loads acting on an optical instrument may also change the shape of the optical surfaces as illustrated in Fig. 4.5. Peak-to-valley (p-v) and root-mean-square (rms) values are typically used to quantify a discrete set of surface displacements. The relationship between p-v and rms is dependent upon the deformed shape as shown for select examples in Fig. 4.6.

4.2.1 Optical surface deformations and wavefront error

As an optical surface deforms under a mechanical load, wavefront error is introduced into the optical system. The relationship between surface error and wavefront error for a refractive surface is given by

$$WFE = (n\cos\theta - n'\cos\theta')SE, \tag{4.1}$$

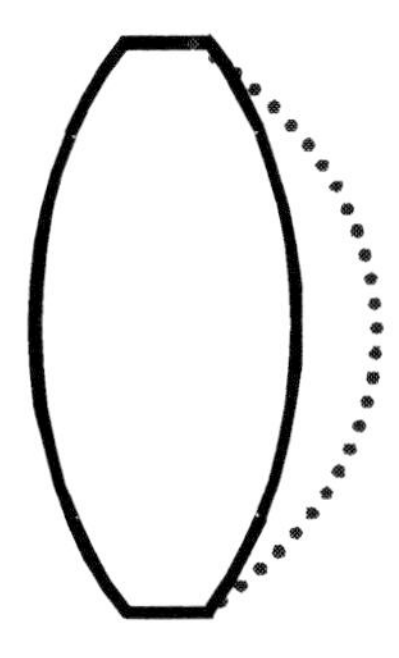

Figure 4.5 Mechanical loads may deform the shape of an optical element.

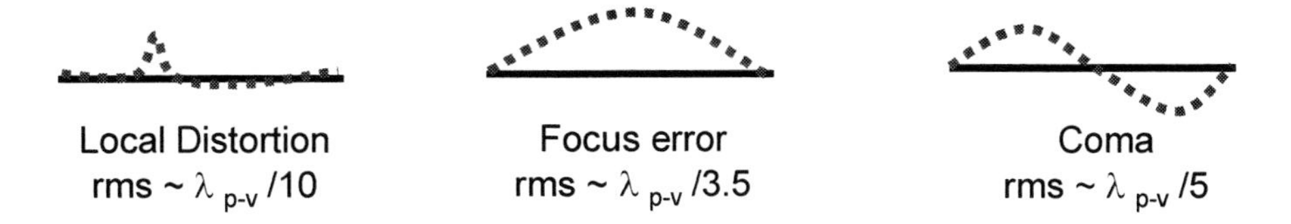

Figure 4.6 Relationship between p-v and rms surface error is dependent on the deformed surface shape.

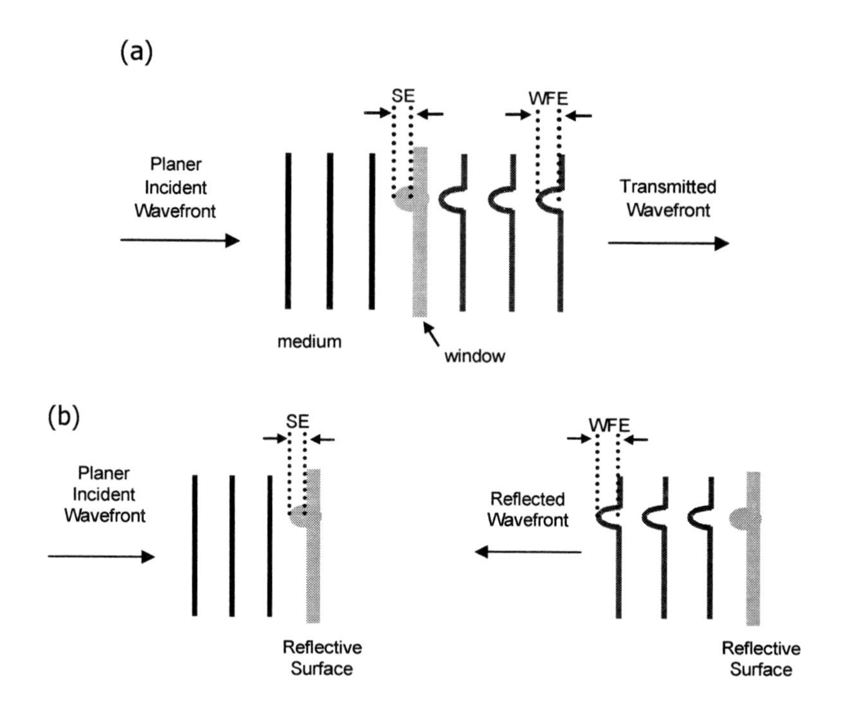

Figure 4.7 (a) Wavefront error from a bump on a refractive surface, (b) wavefront error from a bump on a reflective surface.

where SE represents the magnitude of the surface error, n represents the index of refraction of the medium, n' is the index of the optical element, θ is the angle of incidence, and θ' is the angle of refraction. For a ray traveling at normal incidence, the wavefront error simplifies to

$$\text{WFE} = (n - n')SE. \tag{4.2}$$

This is depicted schematically in Figure 4.7(a) where it is assumed that the index of the window is greater than the index of the medium.

The wavefront error for a wavefront reflecting off a deformed optical surface is given by

$$\text{WFE} = 2SE\cos\theta, \tag{4.3}$$

where θ is the angle of incidence. For a wavefront at normal incidence, as depicted in Figure 4.7(b), this simplifies to

$$\text{WFE} = 2SE. \tag{4.4}$$

In computing the wavefront error for a ray reflecting off a deformed surface, it is often incorrectly assumed that the reference plane from which the OPD is measured is parallel to the undeformed surface. The correct OPD is computed using a reference plane normal to the reflected rays as shown in Fig. 4.8.

4.2.2 Surface normal deformations

Deformations normal to the optical surface are typically used in interferogram files to describe the change in shape of an optical surface. Surface normal displacements are shown in Fig. 4.9 and may be computed for each point on a given surface by the dot product of the finite element displacement vector *(dx, dy, dz)* with the surface normal vector. For a spherical surface, the surface normal displacement, d_{sn}, at a given (x, y) position, and surface curvature, ρ, may be computed using the following relationship:

$$d_{sn} = dz\sqrt{1-\rho^2(x^2+y^2)} - \rho\,(xdx+ydy). \tag{4.5}$$

It is assumed here that the z-axis is parallel to the optical axis.

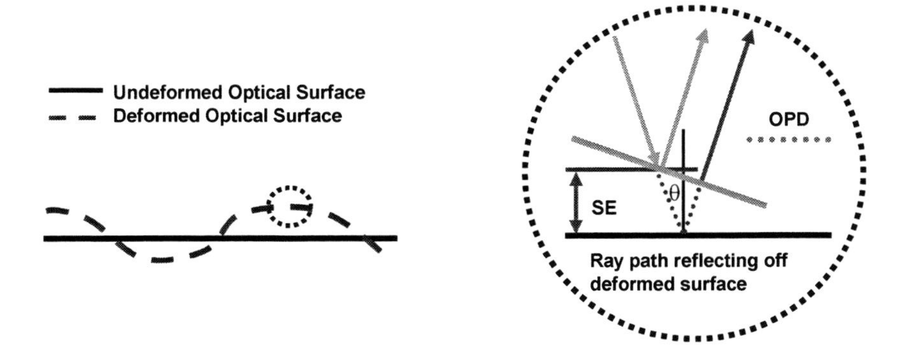

Figure 4.8 OPD from a deformed reflective surface.

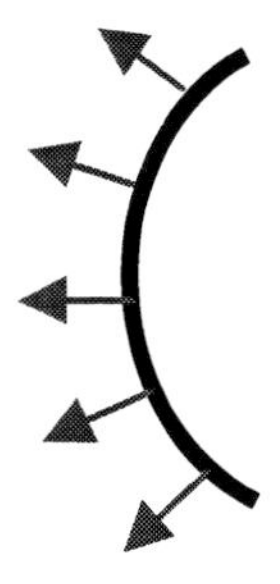

Figure 4.9 Surface normal displacements.

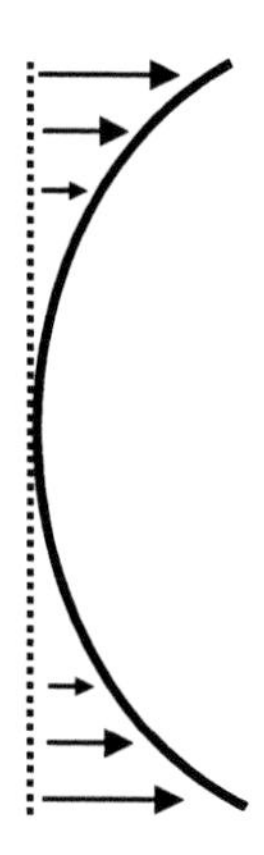

Figure 4.10 Sag displacements.

4.2.3 Sag deformations

The shape of an optical surface is typically defined by the sag of the surface as a function of radial position as shown in Fig. 4.10. Thus, perturbations to the surface shape may be represented by changes in the sag. In general, the sag deformation is not equal to the finite element computed z-displacement since the node position may also be laterally displaced. This is shown for an optical surface supported at the vertex undergoing a uniform increase in temperature in Fig. 4.11. The temperature increase causes the radius of curvature to increase; thus, the sag value for any position on the optical surface is negative. However, the z-displacement as computed by the finite element model is positive. The sag change, ds, may be computed for small perturbations (dx, dy, dz) using the following equation[3]:

$$ds = dz - \frac{\partial z(r_o)}{\partial r}\sqrt{dx^2 + dy^2}. \tag{4.6}$$

The sag value equals the z-displacement only when the node is not displaced in the radial direction. Under gravity loading, for instance, the finite element computed z-displacement closely approximates the sag value.

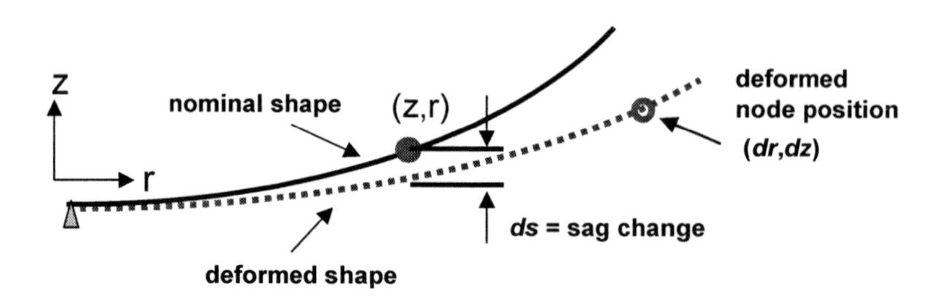

Figure 4.11 FEA z-displacement vs. sag displacement.

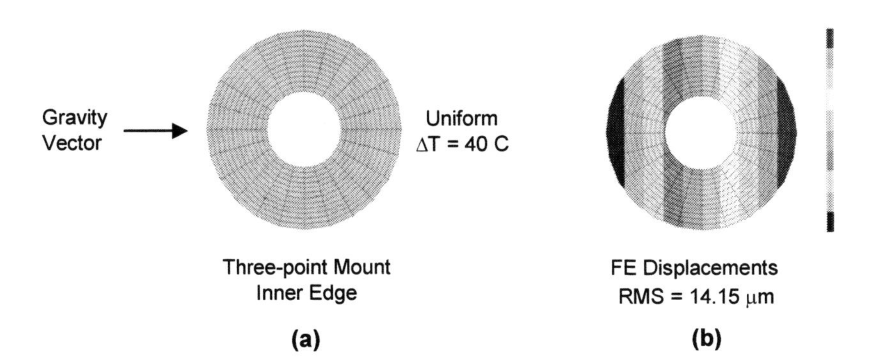

Figure 4.12 (a) Gravity and thermal loads acting on a primary mirror, (b) resulting surface deformations.

4.2.4 Optical surface deformations and Zernike polynomials

Zernike polynomials offer several advantages in representing a discrete set of optical surface deformations, whether they are surface normal or sag displacements. First they help facilitate insight into the behavior and performance of the optical system by identifying the primary shapes of the deformed surface. Second, individual polynomial terms may be removed from the data without affecting the value of the other terms (assumes the Zernike polynomials are orthogonal). And third, the Zernike terms provide a method to help verify the finite element model.

The following example discusses the use of Zernike polynomials to represent the sag deformations of a primary mirror of a Cassegrain telescope. The telescope is subject to gravity acting perpendicular to the optical axis and a uniform temperature change of 40°C. The finite element model and resulting contour map is shown in Fig. 4.12. The magnitude and phase of the Zernike terms and the residual rms and p-v error of the Zernike fit is listed in units of visible waves in Table 4.1.

The two dominant Zernike terms representing the deformed surface are tilt and focus, which mask the higher-order surface deformations as illustrated in Fig. 4.13(a) and 4.13(b). An advantage of using Zernike polynomials is that these terms may be removed from the data without affecting the contributions of the other terms. (In this example, we are assuming the loss in orthogonality of the Zernike terms due to the central hole is insignificant). For optical systems with active control or compensating elements that can correct the rigid-body and/or focus errors of a given surface, the higher-order terms represent the surface error responsible for the loss in image quality. Design modifications may then concentrate on minimizing these higher-order surface errors. The residual values indicate how well the Zernike terms fit the actual data. In this example, the residual error is a small fraction of the actual data indicative of an accurate fit, as shown in Fig. 4.13(c).

Table 4.1 Optical surface deformations represented by Zernike polynomials.

ORDER TERM	N	M	ABERRATION TYPE	MAGNITUDE (WAVES)	PHASE (DEG)	RESIDUAL RMS	RESIDUAL P-V
			Input Surface Data			14.15	48.7
1	0	0	Piston	3.07	0	13.65	48.7
2	1	1	Tilt	25.09	90	1.88	7.1
3	2	0	Focus	3.87	0	0.52	2.3
4	2	2	Pri Astigmatism	0	–90	0.52	2.3
5	3	1	Pri Coma	0	–90	0.52	2.3
6	4	0	Pri Spherical	–0.81	0	0.41	1.6
7	3	3	Pri Trefoil	0.99	30	0.16	0.7
8	4	2	Sec Astigmatism	0	0	0.16	0.7
9	5	1	Sec Coma	0	90	0.16	0.7
10	6	0	Sec Spherical	0.21	0	0.14	0.6
11	4	4	Pri Tetrafoil	0	0	0.14	0.6
12	5	3	Sec Trefoil	0.39	–30	0.07	0.4
13	6	2	Ter Astigmatism	0	–90	0.07	0.4
14	7	1	Ter Coma	0	–90	0.07	0.4
15	8	0	Ter Spherical	–0.05	0	0.07	0.4
16	5	5	Pri Pentafoil	0	0	0.07	0.4
17	6	4	Sec Tetrafoil	0	45	0.07	0.4
18	7	3	Ter Trefoil	0.21	30	0.03	0.2
19	8	2	Qua Astigmatism	0	0	0.03	0.2
20	9	1	Qua Coma	0	90	0.03	0.2
21	10	0	Qua Spherical	0	0	0.03	0.2
22	12	0	Qin Spherical	0	0	0.03	0.2

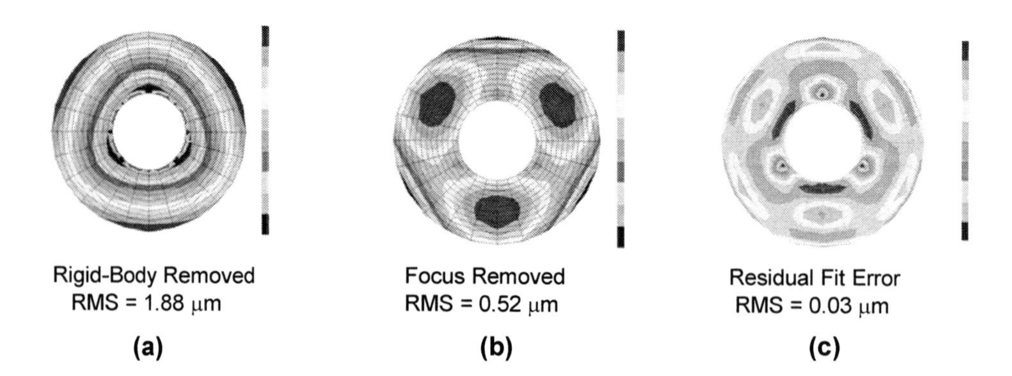

Figure 4.13 (a) Rigid-body Zernike terms removed, (b) rigid-body and focus Zernike terms removed and (c) residual error plot showing how well the Zernike polynomials fit the data.

Since the Zernike polynomials are fit to a single vector quantity (surface normal or sag data), they may not represent the rigid-body motion of an optical surface in six degrees of freedom. For instance, fitting Zernike terms to the surface displacements of a flat optical surface yields no information about whether the surface was laterally displaced or rotated about the optical axis. When computing optical element errors, it is common practice to remove the rigid-body terms (see next section) and represent the higher-order surface deformations with Zernike polynomials.

4.2.5 Computing rigid-body motions from optical surface deformations

For optical surfaces that are represented by a grid of nodes, the rigid-body motion of the surface (three translations, T_x, T_y, T_z, and three rotations, R_x, R_y, R_z) may be computed as the area-weighted average motion. The displacements at node position x_i, y_i, and z_i, due to rigid-body motions is defined by $d\tilde{x}_i$, $d\tilde{y}_i$, and $d\tilde{z}_i$, and are expressed as

$$\begin{aligned} d\tilde{x}_i &= T_x + z_i R_y - y_i R_z \\ d\tilde{y}_i &= T_y - z_i R_x + x_i R_z \, . \\ d\tilde{z}_i &= T_z + y_i R_x - x_i R_y \end{aligned} \tag{4.7}$$

A least-squares fit may be used to compute the average rigid-body motions of a given surface by defining the error, E, as the difference between the actual optical surface nodal displacements, dx_i, dy_i, and ds_i, and the rigid-body nodal displacements, $d\tilde{x}_i$, $d\tilde{y}_i$, and $d\tilde{z}_i$:

$$E = \sum_i w_i \left[(dx_i - d\tilde{x}_i)^2 + (dy_i - d\tilde{y}_i)^2 + (ds_i - d\tilde{z}_i)^2 \right]. \tag{4.8}$$

Note that the sag displacement (ds) is used in these calculations. The best-fit motions are found by taking partial derivatives with respect to each term and setting the result to zero. For example, the resulting equation for translation in the x-direction is shown below:

$$\sum_i w_i T_x + \sum_i w_i z_i R_y - \sum_i w_i y_i R_z = \sum_i w_i dx_i . \tag{4.9}$$

Repeating this for the six rigid-body equations results in six simultaneous equations to solve.

4.2.6 Representing shape changes in the optical model

4.2.6.1 Polynomial surface definition

Representing a deformed optical surface in the optical model may be accomplished by fitting the sag displacements to a polynomial series, and by defining the optical surface using a polynomial surface definition. For example, the finite element derived sag deformations may be represented by Zernike coefficients as perturbations to the base surface shown below:

$$s = \frac{cr^2}{1+\sqrt{1-(1+k)c^2r^2}} + \sum a_i Z_i, \tag{4.10}$$

where s is the sag of the optical surface, the first term is the nominal surface definition, a_i are the Zernike coefficients, and Z_i are the Zernike polynomials. Optical design codes offer an array of surface definitions, including the use of Zernike polynomials, x-y polynomials, even and odd power polynomial aspheres, and anamorphic aspheres with bilateral symmetry.

4.2.6.2 Grid sag surface

A surface definition offered by ZEMAX is the grid sag surface. This surface has a shape defined by a base plane, sphere, conic asphere, or polynomial plus additional sag terms defined by a rectangular array of sag values as defined below:

$$s_{ij} = s_{base} + s(x_i, y_j). \tag{4.11}$$

This format requires the sag value, ds, and the first derivatives in the x and y-directions, $\partial(ds)/\partial x$, $\partial(ds)/\partial y$, along with the cross-derivative, $\partial(ds)^2/\partial x\partial y$. ZEMAX determines the surface shape by a bicubic spline interpolation of the sag values. The rotation values help ensure a smooth fit over the boundary points. A finite difference method is used to estimate the derivatives if the rotation values are not supplied by the user.

4.2.6.3 Interferogram files

Interferogram files provide another method to represent surface deformations in the optical model. This file format was originally used to account for optical test data and was expanded to represent externally derived data. The data in an interferogram file is assumed normal to the optical surface and may be represented by Zernike polynomials or a uniform rectangular array. CODE V, for example, uses surface interferogram files to represent small perturbations in refractive and reflective surfaces. Other types of interferogram files are used in CODE V including wavefront and stress birefringence interferogram files, which are discussed in later sections.

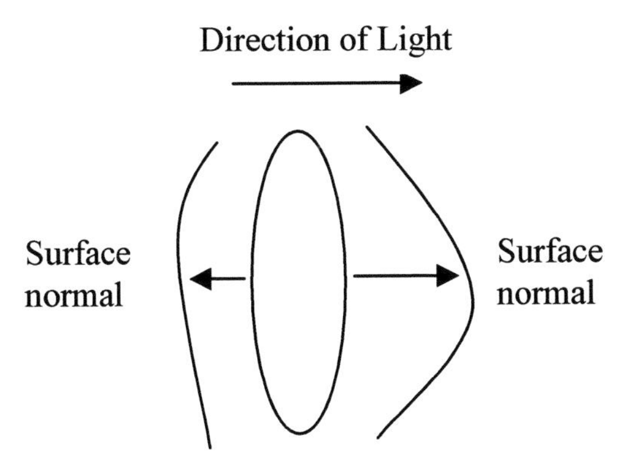

Figure 4.14 Sign convention for CODE V surface interferogram files.

Ray deviations and OPD are computed based on the interferogram data along with the indices of refraction and angle of incidence for a given ray. Use of the Zernike format provides a more accurate representation if an accurate fit is achieved. Surface deformations and slope data is computed directly from the polynomial representation. The grid format is useful to represent high-frequency surface errors, such as edge roll-off and localized mounting effects, when an accurate Zernike fit cannot be achieved. In this case, for rays that do not intersect a data point in the uniform grid, the surface deformation and slope is interpolated or taken as the nearest neighbor.

As with assigning rigid-body perturbations to an optical surface, understanding and relating the finite element coordinate system to the optical surface coordinate system is critical for a successful representation. For instance, in CODE V, a positive surface deformation represents a "bump" on the optical surface, as shown in Fig 4.14. This is consistent with measuring the surface from the "air" side of the element.

In addition, it is critical to align and place the interferogram file at the correct location and with the proper orientation on the optical surface. In general, commands are available to scale, mirror, (reverse or flip it), rotate, and decenter the interferogram file to the correct position. Test cases should always to be run to verify that the position and orientation of the interferogram files are correct.

4.2.6.4 Interpolation

In general, creating a grid interferogram file or grid sag data requires the surface displacements computed at the finite element grid points to be interpolated to a uniform grid as shown in Fig. 4.15. The accuracy of the interpolation method is critical for high-performance optical systems, specifically near mounting areas or areas of rapidly varying displacements. One method to interpolate surface data to a uniform grid is using Delaunay triangulation techniques including nearest neighbor, linear, and cubic. Another method to interpolate data to a uniform grid is to use the finite element shape functions.[4,5] In this approach, values are interpolated to the grid points using the shape functions from the surface element in which the grid point falls. For 3D models, this method requires a set of "dummy" plate elements to be modeled on the optical surface. The structural thickness of this coating may be made arbitrarily small. An example of interpolating a finite element mesh to a uniform grid is shown for Delaunay triangulation techniques (nearest neighbor and cubic) and cubic finite element shape functions in Fig. 4.16.

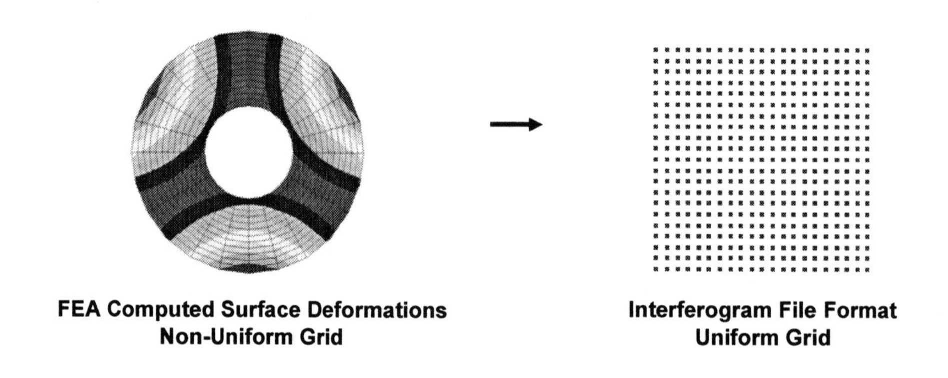

Figure 4.15 Interpolating FEA displacements to a uniform grid.

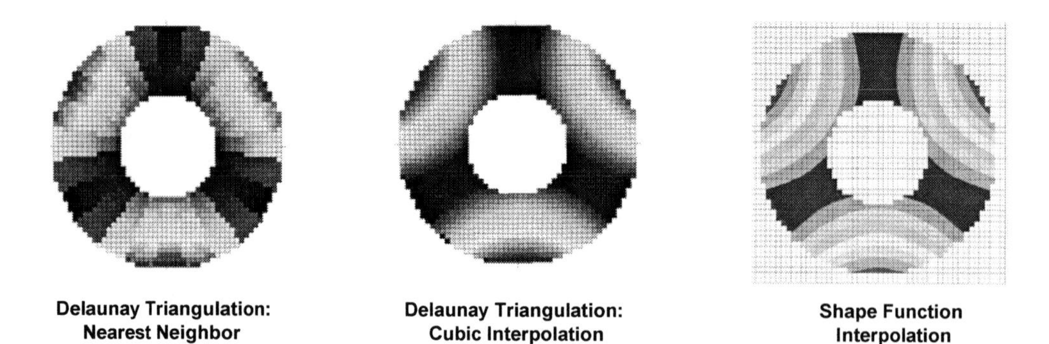

Figure 4.16 Interpolation using Delaunay triangulation and FE shape functions.

4.2.7 Focus, Zernikes, and radius of curvature

Computing the change in the radius of curvature of a deformed optical surface may be approximated using the Zernike focus term that has been fit to the sag deformations of the optical surface. This is an approximation because higher-order terms are required to define the sag, s, of a true spherical surface as shown in the series expansion below:

$$s = \frac{r^2}{2R} + \frac{r^4}{8R^3} + \frac{r^6}{16R^5} + \frac{r^8}{128R^7} + \ldots, \tag{4.12}$$

where r is the radial extent of the surface, and R is the radius of curvature of the optical surface as illustrated in Fig. 4.17. This approximation may be demonstrated by fitting Zernike polynomials to an optical surface with a pure radius of curvature change. The focus term and the higher-order rotationally symmetric terms are used to describe the deformed shape. For optical surfaces that are not highly curved, the sag contribution of a spherical surface is dominated by the parabolic variation given by the first term in Eq. (4.12). The focus term is a "best-fit" quadratic to the deformed shape and may be used to estimate a change in the radius of curvature, ΔR, of an optical surface as given below:

$$\Delta R = 4A_{20}\left(\frac{R}{r}\right)^2, \tag{4.13}$$

where A_{20} is the coefficient of the Zernike focus term.

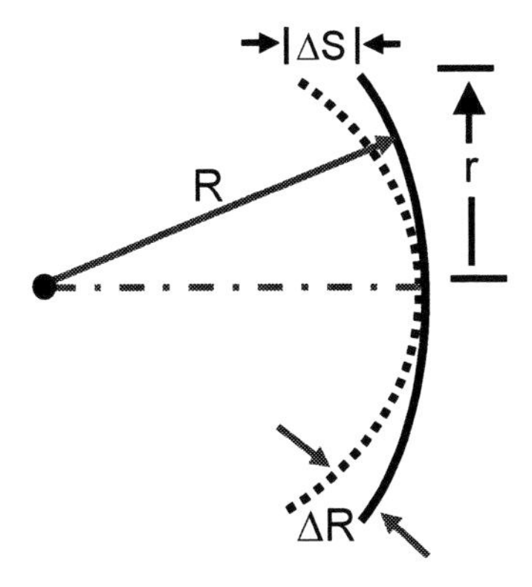

Figure 4.17 Radius of curvature change.

4.2.8 Finite element derived spot diagrams

Optical surface quality may be evaluated using a spot diagram computed directly from the finite element model.[6,7] Here, collimated light is assumed incident on a finite element surface. Rays are modeled as rigid bars from each node on the optical surface to unmerged nodes located at the image point. The image point is located at twice the focal length since the angle of the reflected rays off the deformed optical surface is twice the angle error as computed by the finite element model. This computes the correct ray displacement. A plot of the location of the displaced nodes on the image plane gives the spot diagram. A corresponding rms spot size may then be computed.

4.3 Line-of-Sight Jitter

Line-of-sight (LOS) jitter is the time-varying motion of the image on the detector plane, which may be caused by internal or external dynamic loads acting on an optical system. For example, airborne and vehicular mounted systems are typically subjected to external vibratory loads during operation. Jitter may also be caused by internal effects in an optical instrument due to friction variations, mounting imbalances in moving components, or from the slewing of their tracking systems.

Image jitter degrades optical performance by increasing the blur diameter of the PSF. This results in blurring of the object and loss in resolution as illustrated in Fig. 4.18. A corresponding MTF curve may be computed by considering only the effects of image motion. This curve may then be multiplied by the nominal MTF curve of the optical system to compute a system MTF.

The magnitude of image blur depends on the amplitude and frequency of the mechanical disturbance, and the exposure/integration time of the detector. Exposure time relates to the use of photographic film and integration time to the photon-collecting time of a pixel array. The acceptable amount of image motion

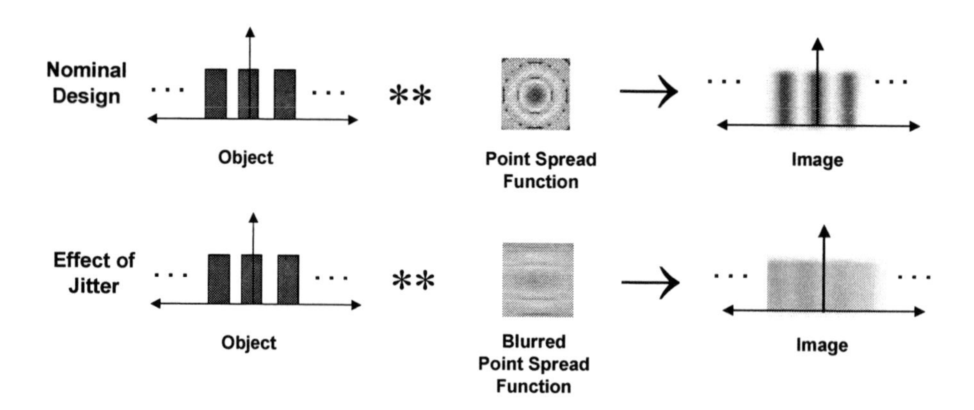

Figure 4.18 Image formation including the effects of jitter.

depends on the application. For instance, a rule of thumb LOS jitter requirement for IR sensors is to minimize the image motion to a tenth of a pixel.[8]

In this section, we are considering transverse image motion or motion in the plane of the detector. However, vibratory loads may also induce longitudinal image motion or defocus that displaces the image along the optical axis. The influence of longitudinal image motion on optical performance is typically not nearly as severe as transverse image motion of the same amplitude. Furthermore, computing the effect of longitudinal image motion on optical system performance is more complicated than for transverse image motion since the transfer functions are coupled.[9]

4.3.1 Computing image motion

Image motion, as a function of time and frequency, may be computed using finite element analysis. This method assumes that the image motion is a linear function of the optical element displacements. The position of the image will be defined as the position of the on-axis chief ray. (The motion of the image centroid may also be used). The first step in the analysis is to compute the optical sensitivities typically performed by a tolerancing algorithm of an optical design code. The optical sensitivity coefficients relate the motion of each of the six degrees-of-freedom of the optical element to the motion of the on-axis chief ray. Image motion is then computed in the finite element model using multipoint constraint equations where the weighting coefficients are the optical sensitivity coefficients.

Jitter may be computed in either object or image space. In image space, image jitter is computed as the lateral motion of the image, Δ_x and Δ_y, on the detector plane. In object space, angular units, θ_{ele} and θ_{azi}, may be used to describe the change in the elevation and azimuthal angle of the object. For an object at infinity, the relationship between jitter in image and object space is shown in Figure 4.19 and expressed below:

$$\theta_{\text{obj}} = \frac{\Delta_{\text{image}}}{f_{\text{eff}}}. \tag{4.14}$$

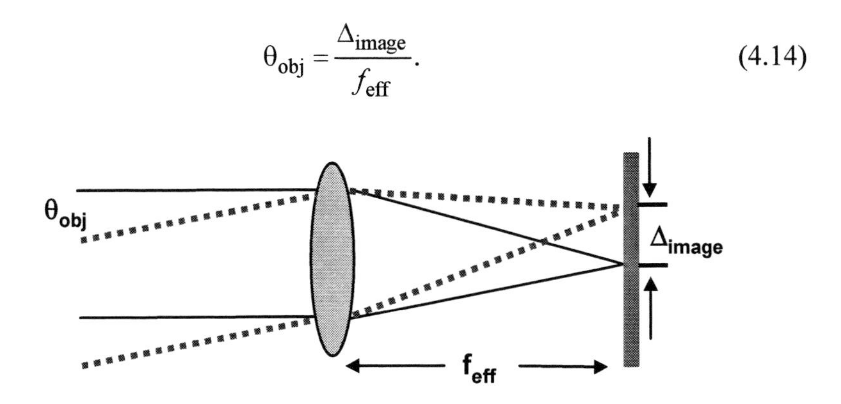

Figure 4.19 Jitter in object and image space for an object at infinity.

Optical sensitivities are computed accordingly to relate optical element rigid-body motions to the proper space and are denoted by the matrices $[L]_{obj}$ and $[L]_{img}$. The two equations relating the rigid-body motions for image and object space, respectively, are given as

$$\begin{Bmatrix} \Delta_x \\ \Delta_y \end{Bmatrix} = [L]_{img} \{X\}_{optics}, \tag{4.15}$$

and

$$\begin{Bmatrix} \theta_{ele} \\ \theta_{azi} \end{Bmatrix} = [L]_{obj} \{X\}_{optics}. \tag{4.16}$$

The rigid-body motions in six degrees-of-freedom are represented by the vector $\{X\}_{optics}$.

In employing this modeling technique to predict LOS jitter, the optical perturbations computed in the optical design code must be properly related to the mechanical perturbations computed by the finite element model.

4.3.2 Example: Cassegrain telescope

Line-of-sight jitter equations are developed for a Cassegrain telescope as shown in Fig. 4.20. The optical elements include the primary mirror, M_1, secondary mirror, M_2, and lens, L_1.

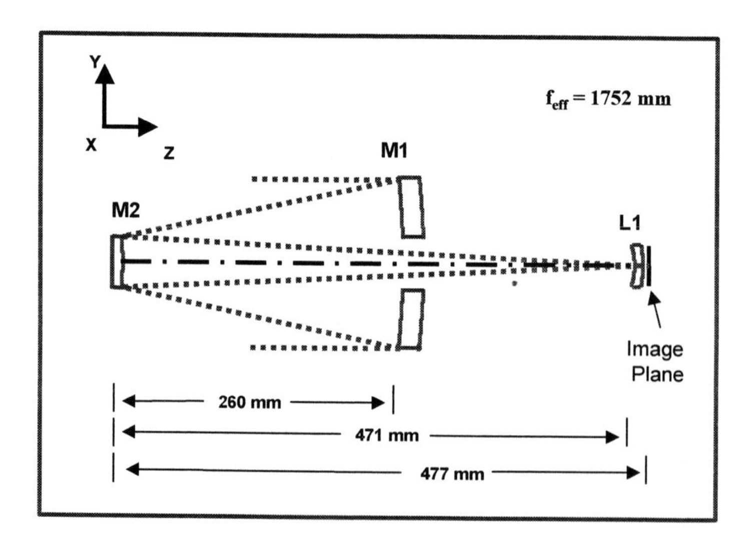

Figure 4.20 Cassegrain telescope.

Image space optical sensitivities are computed for each of the optical elements in six degrees of freedom and are listed in Table 4.2. The equations for the image motion, Δ_x and Δ_y, are given as

$$\Delta_x = 4.72M_1\Delta x - 3.5M_1\Delta\beta - 3.62M_2\Delta x + 1.05M_2\Delta\beta - 0.095L_1\Delta x + 0.003L_1\Delta\beta - (\Delta x \text{ of image plane}), \tag{4.17}$$

and

$$\Delta_y = 4.72M_1\Delta y + 3.5M_1\Delta\alpha - 3.62M_2\Delta y - 1.05M_2\Delta\alpha - 0.095L_1\Delta y - 0.003L_1\Delta\alpha - (\Delta y \text{ of image plane}). \tag{4.18}$$

Rigid-body checks are important to verify the accuracy of the LOS jitter equations. A couple of examples are presented below. First, the object, telescope, and detector are decentered in the Y-direction by 1 μm. Substituting the appropriate values into Eqs. (4.17) and (4.18) yields the image motion:

$$\Delta_x = 0, \tag{4.19}$$

and

$$\Delta_y = 4.72(1.0) + 3.5(0.0) - 3.62(1.0) - 1.05(0.0) - 0.095(1.0) - 0.003(0.0) - 1.0 = 0.0. \tag{4.20}$$

Table 4.2 Cassegrain telescope sensitivity matrix.

PERTURBATION	SENSITIVITY	PERTURBATION	SENSITIVITY
M1Δx	4.720	M2Δα	–1.050
M1Δy	4.720	M2Δβ	1.050
M1Δz	0.000	M2Δγ	0.000
M1Δα	3.500	L1Δx	–0.095
M1Δβ	–3.500	L1Δy	–0.095
M1Δγ	0.000	L1Δz	0.000
M2Δx	–3.620	L1Δα	–0.003
M2Δy	–3.620	L1Δβ	0.003
M2Δz	0.000	L1Δγ	0.000

As expected, the position of the image does not change. A second equation check is to apply a rigid-body rotation to the optical instrument. A rotation about the vertex of the secondary mirror θ_{yz} of –1 μrad is applied. Substituting in the resulting mechanical perturbations of the individual optical elements results in the following expressions:

$$\Delta_x = 0, \tag{4.21}$$

and

$$\Delta_y = 4.72(0.26) + 3.5(-1.0) - 3.62(0.0) - 1.05(-1.0) - 0.095(0.471) - 0.003(-1.0) - 1.0(0.477) = -1.75\ \mu\text{m}. \tag{4.22}$$

The image motion, Δ_y, may be represented as a change in the angular position of the object using Eq. (4.14):

$$\theta_{\text{obj}} = \frac{\Delta_{\text{image}}}{f_{\text{eff}}} = -1.75\ \mu\text{m} / 1750\ \text{mm} = -1\ \mu\text{rad}\,. \tag{4.23}$$

Thus, the change in the angle of the object is equal to the rotation of the optical instrument.

Multi-point constraint equations may be written to represent the Eqs. (4.17) and (4.18) within the finite element code. This allows the change in image position to be computed as a function of time or frequency. An example using this technique to compute the image motion for an airborne sensor is given in Ref. 10. Design efforts centered on minimizing the line-of-sight jitter to a quarter of the pixel size over the integration time of the detector.

4.3.3 Quantifying the effects of jitter using the MTF

An overall optical system MTF may be computed accounting for the effects of jitter by multiplying the nominal optical system MTF by the MTF of the optical system due solely to image jitter. This calculation may be performed by optical design software by entering the magnitude and direction of the image motion as shown in Fig. 4.21. Here the MTF is shown for an airborne camera operating on the ground, with no image jitter, and in-flight with image jitter. Closed-form expressions also exist to compute the MTF at a specific spatial frequency for constant velocity and sinusoidal image motion.[11] This value may be multiplied by the nominal optical system MTF to compute the overall optical system MTF value for that spatial frequency. This can be repeated for spatial frequencies of interest.

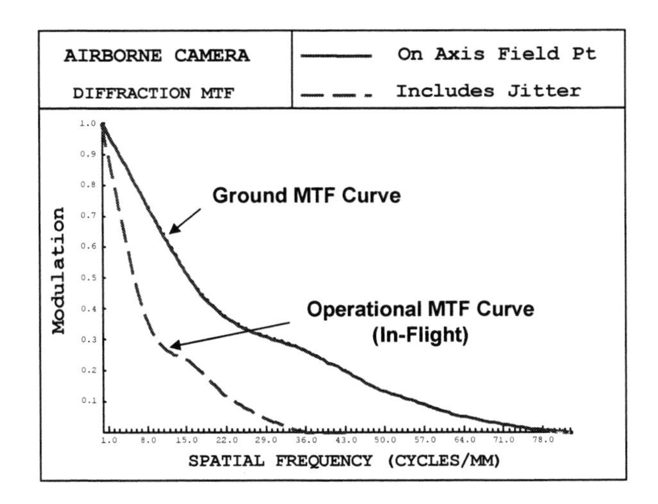

Figure 4.21 Effect jitter has on the MTF.

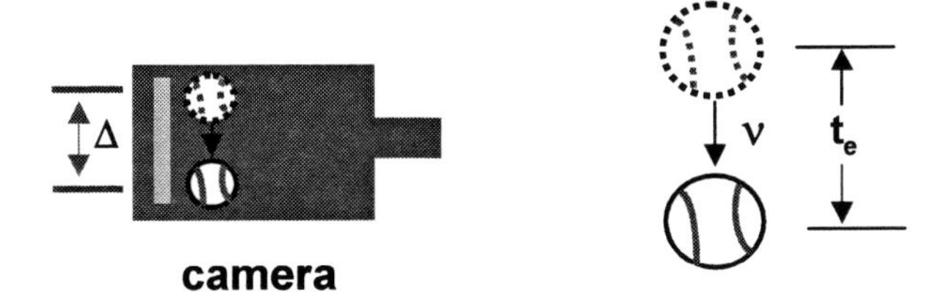

Figure 4.22 Constant velocity image motion.

4.3.3.1 Constant velocity image motion

Consider a camera taking a snapshot of a baseball moving at constant velocity as illustrated in Fig. 4.22. The image velocity, v, times the exposure or integration time, τ_e, yields the image displacement, Δ_{cv}. The MTF is computed as

$$\mathrm{MTF}(\xi) = \frac{\sin \pi\xi\Delta_{cv}}{\pi\xi\Delta_{cv}} , \tag{4.24}$$

where ξ is the spatial frequency in cycles per millimeter.

4.3.3.2 High-frequency sinusoidal image motion

For sinusoidal image motion characteristic of airborne and vehicular instruments, the loss in resolution is dependent upon the period of the exposure/integration time, t_e, to the period of the jitter, t_j. High-frequency jitter occurs when the jitter

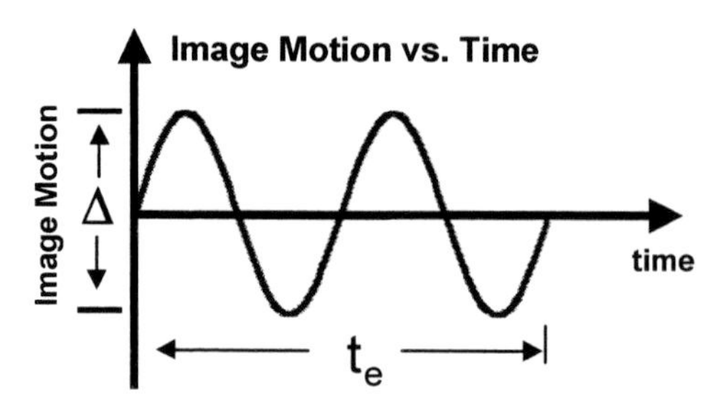

Figure 4.23 High-frequency sinusoidal image motion.

frequency is greater than the exposure/integration frequency, or when $t_e > t_j$. This allows the detector to integrate over several cycles of image displacement as illustrated in Fig. 4.23. The range of the sinusoidal image displacement gives the blur diameter, Δ_r. The corresponding MTF is computed below:

$$\mathrm{MTF}(\xi) = J_o(2\pi\xi\Delta_r)\,, \tag{4.25}$$

where J_o is the zero-order Bessel function.

4.3.3.3 Low-frequency sinusoidal image motion

For low-frequency jitter, i.e., when the jitter frequency is less than the exposure frequency or $t_e < t_j$, the magnitude of the image motion depends upon the phasing between the exposure frequency and the jitter frequency. Image displacement for sinusoidal motion as a function of time is shown in Fig. 4.24. The image displacement depends upon the position of the image at the start of the exposure period and the duration of the exposure. The maximum and minimum image motion for an exposure time of, t_e, is given by

$$\Delta_{\min} = A\left\{1-\cos\left[\left(\frac{2\pi}{T}\right)\left(\frac{t_e}{2}\right)\right]\right\}, \tag{4.26}$$

and

$$\Delta_{\max} = 2A\left[\left(\frac{2\pi}{T}\right)\left(\frac{t_e}{2}\right)\right]. \tag{4.27}$$

The MTF may be bounded using the image motion computed in Eqs. (4.26) and (4.27) and substituted in Eq. (4.24).

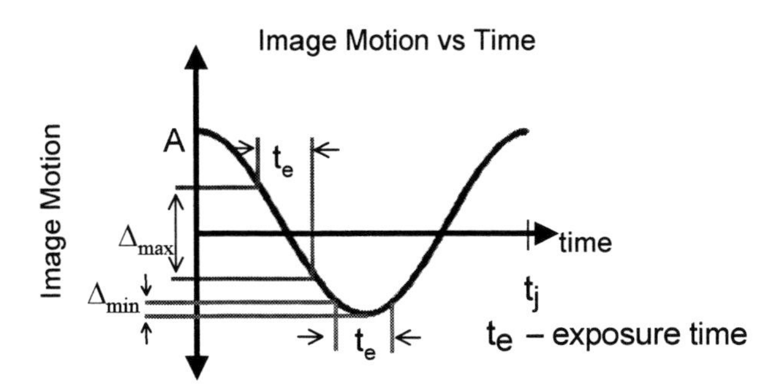

Figure 4.24 Low-frequency sinusoidal image motion.

4.3.3.4 Image jitter and the MTF for an airborne camera

The MTF loss for an airborne camera system is computed at a spatial frequency of 20 cycles/mm. The image is vibrating at the optical instrument's fundamental frequency of 150 Hz with a total displacement range of 7 μm as illustrated in Fig. 4.25. The integration time of the detector is 25 msec. This yields an integration frequency of 40 Hz. Since the jitter frequency (150 Hz) is greater than the integration frequency (40 Hz), Eq. (4.25) may be used to compute the MTF loss as given below:

$$\text{MTF}(20) = J_o[(2\pi)(20)(0.007)] \approx 0.82\,. \tag{4.28}$$

This value may be multiplied by the nominal optical system MTF to yield the total system MTF for the given spatial frequency.

4.3.4 Control system interaction

Servo control loops are often used in tracking systems to provide pointing control. The control system may help correct rigid-body image motion at frequencies below the control system bandwidth. This interaction may be modeled within the finite element code. For instance, in MSC/Nastran, control system effects are modeled as an adjunct system to the mechanical structure using extra points and transfer functions.[12]

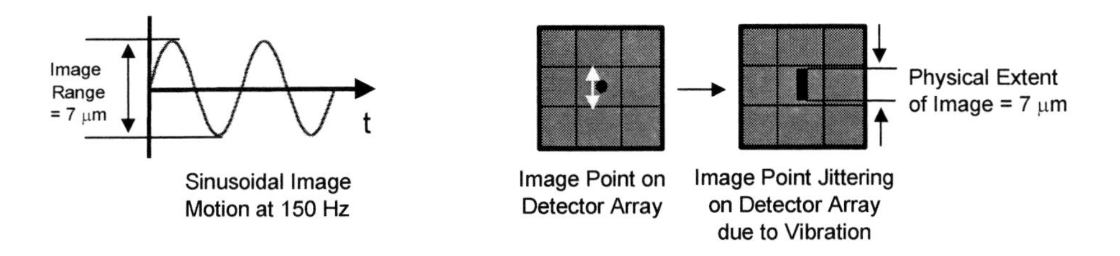

Figure 4.25 Example of sinusoidal image motion.

4.4 Stress Birefringence

Lens elements subject to mechanical stress experience index of refraction changes due to the phenomenon known as the photo-elastic effect or stress birefringence, resulting in wavefront and polarization errors in the optical system. This is illustrated for a glass plate under uniaxial stress in Fig. 4.26. The applied stress modifies the indices of refraction in the directions parallel and perpendicular to the direction of the stress. For a general triaxial state of stress varying within a lens element, the optical properties typically become anisotropic and inhomogenous.

Stress birefringence is an issue for many types of optical systems, including systems for optical lithography, data storage, high-energy lasers, LCD projectors, and telecommunications. For these types of optical systems, integrated modeling techniques help enable design trades of optical performance as a function of glass type and mounting methods. An example of the effects of stress birefringence is demonstrated for a telecommunication demultiplexer in Fig. 4.27. On the left side of the figure, rays are shown passing through the finite element model of the lens element for the unstressed and stressed states. On the right side, the polarization pupil maps reveal that the stress field converts the incident linearly polarized light to circular polarized light at points near the edges of the pupil.

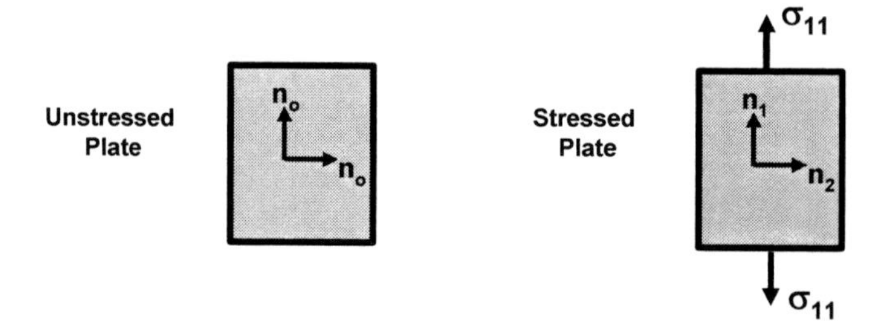

Figure 4.26 Mechanical stress modifies the indices of refraction of transmissive materials.

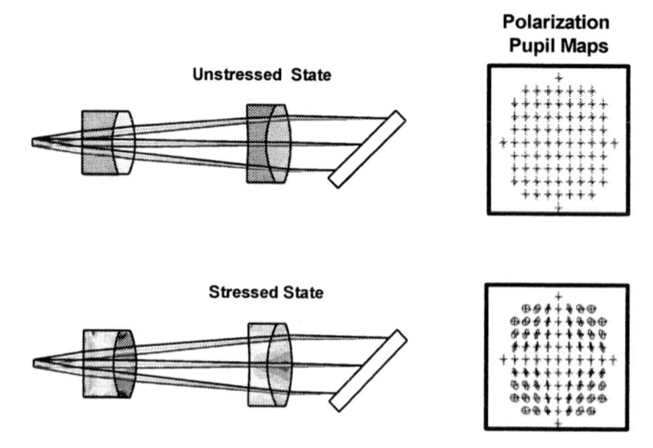

Figure 4.27 Polarization pupil maps representing the effect stress has on the state of polarization for a WDM demultiplexer.

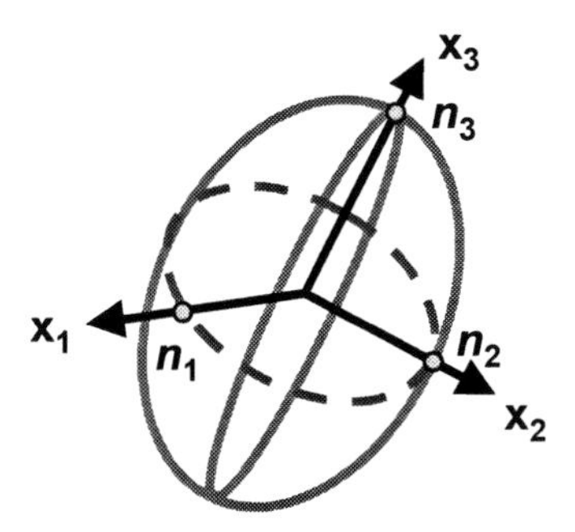

Figure 4.28 Index ellipsoid.

4.4.1 Mechanical stress and the index ellipsoid

The index ellipsoid (also known as the ellipsoid of wave normals, or optical indicatrix) provides a convenient geometrical interpretation of the optical properties of a material by defining the indices of refraction as the semi-axes of the ellipsoid as illustrated in Fig. 4.28. In general, the application of mechanical stress modifies the shape of the index ellipsoid and hence the indices of refraction. The index ellipsoid is defined by a second-degree surface, or quadric, expressed by the following equation:

$$\sum_{ij} \mathrm{B}_{ij} x_i x_j = 1. \tag{4.28}$$

The coefficients of the surface are defined by the dielectric impermeability tensor, which is expressed in matrix form below:

$$\mathrm{B}_{ij} = \begin{bmatrix} \mathrm{B}_{11} & \mathrm{B}_{12} & \mathrm{B}_{13} \\ \mathrm{B}_{12} & \mathrm{B}_{22} & \mathrm{B}_{23} \\ \mathrm{B}_{13} & \mathrm{B}_{23} & \mathrm{B}_{33} \end{bmatrix}. \tag{4.29}$$

For a ray of arbitrary direction traversing a birefringent material, light propagation is affected by the indices of refraction normal to the ray direction. These indices of refraction may be geometrically constructed as the semi-axes of the ellipse centered at the index ellipsoid origin and normal to the ray direction as shown in Fig. 4.29.

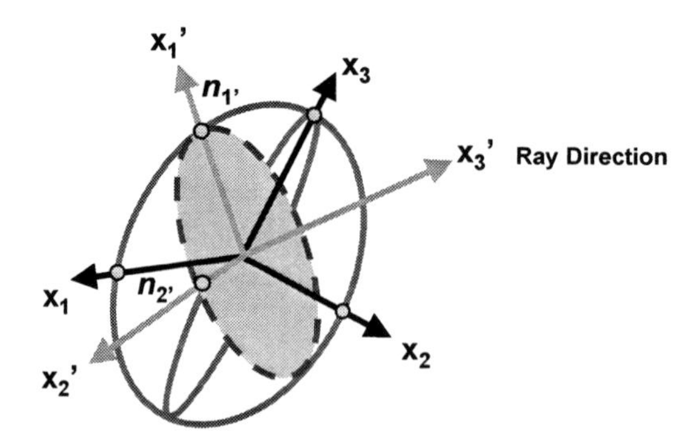

Figure 4.29 Computing the indices of refraction for an arbitrary ray direction.

Mechanical stress changes the indices of refraction by altering the size, shape, and orientation of the index ellipsoid as given by the following fourth-rank tensor transformation:

$$\Delta B_{ij} = q_{ijkl}\sigma_{kl}, \tag{4.30}$$

where q is the stress-optical coefficient matrix, and σ is the stress tensor. The stress-optical coefficient is a material property. For a general crystal structure, this tensor has 36 independent components, whereas for an isotropic material there are only two independent components.

Changes in the dielectric impermeability tensor due to the photo-elastic effect may also be defined using mechanical strain expressed as

$$\Delta B_{ij} = p_{ijrs}\varepsilon_{rs}, \tag{4.31}$$

where p is the elasto-optical coefficient, and ε is the strain tensor.

In general, changes in the dielectric impermeability tensor due to the effects of stress are small and are considered perturbations to the index ellipsoid. Thus, they may be superimposed on the natural birefringence for all crystal systems by adding the terms to the nominal coefficients of the index ellipsoid equation.

4.4.2 Optical errors due to stress birefringence

The change in the dielectric impermeability tensor may be computed for isotropic and crystalline materials using the appropriate stress-optical coefficient matrix. Each material type behaves differently under stress. For example, under a uniform state of stress, isotropic materials become uniaxial. Conversely, various classes of cubic crystals that nominally exhibit isotropic properties may become biaxial. (Uniaxial and biaxial refer to the number of axes in which a given ray traveling parallel to the axis will experience no birefringence). In this section we will develop the equations for isotropic materials.

The stress-optical coefficient matrix for a homogeneous and isotropic material is as follows:

$$q = \begin{bmatrix} q_{11} & q_{12} & q_{12} & 0 & 0 & 0 \\ q_{12} & q_{11} & q_{12} & 0 & 0 & 0 \\ q_{12} & q_{12} & q_{11} & 0 & 0 & 0 \\ 0 & 0 & 0 & q_{44} & 0 & 0 \\ 0 & 0 & 0 & 0 & q_{44} & 0 \\ 0 & 0 & 0 & 0 & 0 & q_{44} \end{bmatrix}, \text{ where } q_{44} = \frac{q_{11} - q_{12}}{2}. \tag{4.32}$$

In the plane normal to the ray direction, the two indices of refraction are defined by the semi-axes of the ellipse centered at the origin of the ellipsoid. When no stress is acting on the material, the ellipse is a circle and there is no birefringence. However, under a mechanical stress defined in an arbitrary *xy* coordinate system:

$$\sigma^o = \begin{bmatrix} \sigma^o_{xx} & \sigma^o_{yy} & \sigma^o_{zz} & \sigma^o_{xy} & \sigma^o_{yz} & \sigma^o_{xz} \end{bmatrix}^T, \tag{4.33}$$

where the z-axis is defined along the ray direction, the material becomes birefringent, and the circle centered at the origin of the ellipsoid becomes an ellipse. The angle, γ, which defines the orientation of the semi-axes of the ellipse, coincides with the direction of the principal stresses, σ_{11} and σ_{22}, in the plane normal to the ray. This angle is computed as

$$\gamma = \frac{1}{2}\tan^{-1}\frac{2\sigma_{xy}}{\sigma_{xx} - \sigma_{yy}}, \tag{4.34}$$

yielding the following stress tensor defined along the in-plane principal stress directions:

$$\sigma = \begin{bmatrix} \sigma_{11} & \sigma_{22} & \sigma_{zz} & 0 & \sigma_{yz} & \sigma_{xz} \end{bmatrix}^T. \tag{4.35}$$

The change in the dielectric impermeability tensor is given by the following:

$$\Delta\beta_{ij} = \begin{bmatrix} q_{11}\sigma_{11} + q_{12}(\sigma_{22}+\sigma_{zz}) \\ q_{11}\sigma_{22} + q_{12}(\sigma_{11}+\sigma_{zz}) \\ q_{11}\sigma_{zz} + q_{12}(\sigma_{11}+\sigma_{22}) \\ 0 \\ q_{44}\sigma_{yz} \\ q_{44}\sigma_{xz} \end{bmatrix}. \tag{4.36}$$

The index changes in the plane normal to the ray direction may be computed by differentiating ΔB and assuming the changes in index are comparatively small, which yields the following:

$$\Delta n_1 = -\frac{1}{2}n_o^3 \Delta B_{11} = -\frac{1}{2}n_o^3\left[q_{11}\sigma_{11} + q_{12}\left(\sigma_{22} + \sigma_{zz}\right)\right], \tag{4.37}$$

and

$$\Delta n_2 = -\frac{1}{2}n_o^3 \Delta B_{22} = -\frac{1}{2}n_o^3\left[q_{11}\sigma_{22} + q_{12}\left(\sigma_{11} + \sigma_{zz}\right)\right], \tag{4.38}$$

where n_o is the nominal index of refraction.

For a wavefront incident on a birefringent material, as illustrated in Fig. 4.30, the incident electric field vector is decomposed into components along the x and y-directions, each traveling paths with different indices of refraction. The difference in the optical path, ΔOPD, between the two ray components is given by the difference in the index of refraction multiplied by the distance the ray traveled, L, as

$$\Delta\text{OPD} = \left(\Delta n_2 - \Delta n_1\right)L = -\frac{1}{2}n_o^3 q\left(\sigma_{11} - \sigma_{22}\right)L, \tag{4.39}$$

where a single stress-optical coefficient, q, is defined as $q = q_{11} - q_{12}$.

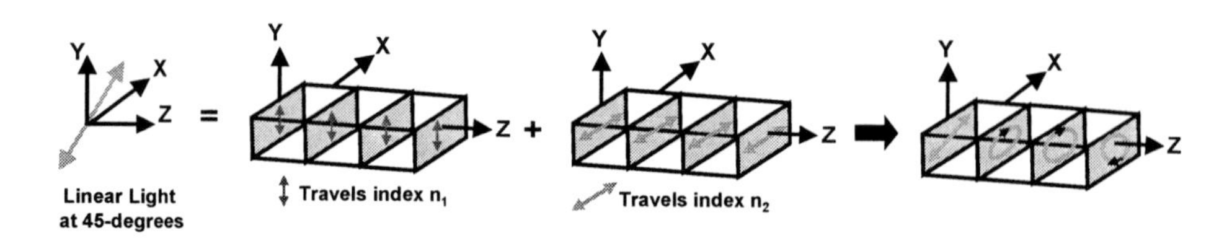

Figure 4.30 For a birefringent material, electric field components travel paths of different indices of refraction resulting in wavefront and polarization errors.

The stress-induced wavefront error is determined by averaging the optical path of the two electric field components:

$$\mathrm{OPD} = \frac{\Delta n_1 + \Delta n_2}{2} L . \tag{4.40}$$

An alternative set of stress-optical coefficients, k_{11} and k_{12}, are often used to express the index changes:

$$\Delta n_1 = k_{11}\sigma_{11} + k_{12}\left(\sigma_{22} + \sigma_{zz}\right), \tag{4.41}$$

and

$$\Delta n_2 = k_{11}\sigma_{22} + k_{12}\left(\sigma_{11} + \sigma_{zz}\right). \tag{4.42}$$

This yields the following OPD difference between the orthogonal components of the electric field:

$$\Delta\mathrm{OPD} = k(\sigma_{11} - \sigma_{22})L, \tag{4.43}$$

where $k = k_{11} - k_{12}$.

4.4.3 Stress-optical coefficients

The stress-optical coefficient is a material property relating the mechanical stress acting on an optical element to the optical properties of the material. The stress-optical coefficients for several Schott glasses at a wavelength of 546 nm are given in Table 4.3.[13] A material that may be resistant to polarization changes under stress may generate significant wavefront error and vice versa. The Schott glasses SF57 and BaK6 are suitable examples. For SF57, the stress-optical coefficient, k, is approximately zero at 546 nm. Thus, minimal changes in polarization occur due to the effects of stress. However, a comparatively large wavefront error may be induced in SF57 due to the large values of k_{11} and k_{12}.

Table 4.3 Stress-optical coefficients for selected Schott glasses.

SCHOTT GLASS STRESS-OPTICAL COEFFICENTS (λ = 546 NM ; UNITS 10^{-6} MM^2/N)			
GLASS	$-K_{11}$	$-K_{12}$	K
FK3	1.0	4.9	3.9
PK2	0.4	3.1	2.7
BK7	0.5	3.3	2.8
BaK6	0.8	3.2	2.4
LaK21	1.0	2.8	1.8
PSK53A	1.5	2.6	1.1
SF1	4.5	6.2	1.7
SF57	6.7	6.7	0.0
SF59	9.0	7.6	–1.4

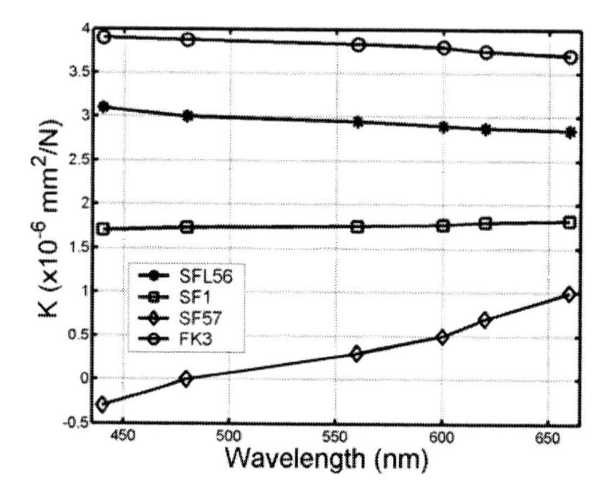

Figure 4.31 Stress-optical coefficient vs. wavelength for selected materials.

Conversely, the glass BaK6 has a comparatively large value of k but relatively small values of k_{11} and k_{12}, which will minimize wavefront error but increase the polarization errors. The stress-optical coefficients are a function of wavelength as shown for selected materials in Fig. 4.31.

4.4.4 Nonuniform stress distributions

For a nonuniform stress field acting on an isotropic material, the magnitude of the indices of refraction, Δn_1 and Δn_2, and orientation, γ, vary at every point along the ray path as illustrated in Fig. 4.32. For an isotropic material, this is equivalent to the ray traversing a series of uniaxial crystals or a series of crossed waveplates of varying birefringence.

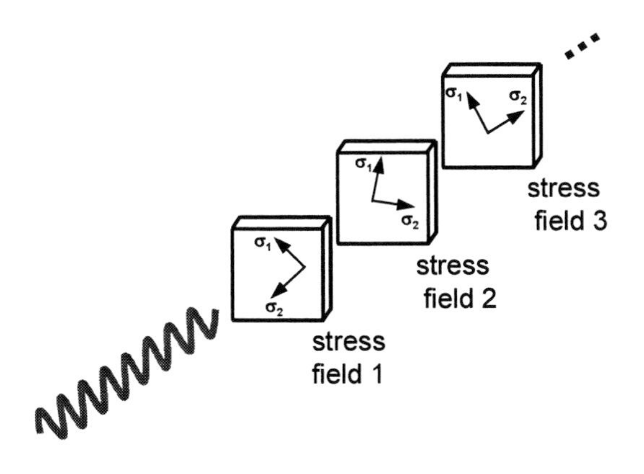

Figure 4.32 Individual ray traversing a nonuniform stress field.

Computing the integrated optical effects of a nonuniform stress distribution may be performed by the use of Jones calculus.[14] At incremental points along the ray path, Jones rotation and retarder matrices are defined. The retarder matrix is used to modify the optical phase of the two orthogonal electric field components as defined below:

$$R(\delta) = \begin{bmatrix} e^{i\delta_1} & 0 \\ 0 & e^{i\delta_2} \end{bmatrix}, \tag{4.44}$$

where the phase change, expressed in radians, is given by

$$\delta_1 = \frac{2\pi \Delta n_1 L}{\lambda} \text{ and } \delta_2 = \frac{2\pi \Delta n_2 L}{\lambda}. \tag{4.45}$$

The rotation matrix is used to rotate between a user-defined coordinate system and the principal coordinate system and is defined as:

$$R(\gamma) = \begin{bmatrix} \cos\gamma & \sin\gamma \\ -\sin\gamma & \cos\gamma \end{bmatrix}. \tag{4.46}$$

A Jones matrix is computed for each incremental stress field, i, using the following relationship:

$$M_i = R(\gamma)_i^T R(\delta)_i R(\gamma)_i , \tag{4.47}$$

and a system-level matrix, M_s, is developed by multiplying each incremental matrix, M_i, given as

$$M_s = M_i \ldots M_2 M_1 . \tag{4.48}$$

The system Jones matrix, M_S, for a given ray defines the integrated effects of the stress field on the optical properties of the material. Changes in the polarization state may be computed using the following expression:

$$E_o = M_s E_i . \tag{4.49}$$

where E_i and E_o represent the input and output Jones vectors. This is depicted schematically in Fig. 4.33. Jones vectors are used to describe the magnitude and phase of the two orthogonal components of the electric field.

The wavefront error may be approximated by averaging the optical path of the two electric field components for each incremental length, L_i, and summed for a given ray path:

$$\mathrm{OPD}_i = \left(\frac{\Delta n_1 + \Delta n_2}{2}\right)_i L_i \rightarrow \mathrm{WFE} = \sum_{i=1}^{n} \mathrm{OPD}_i . \tag{4.50}$$

The two perpendicular components, corresponding to an incident ray, actually travel in different paths within the birefringent medium. The ray deviations due to the effects of stress birefringence are generally considered insignificant, and both ray components are assumed to travel in straight lines and exit at the same point.

4.4.5 Example: stress birefringence

Linear polarized light is incident upon a BK7 window with a uniaxial state of stress as illustrated in Fig. 4.34. The output polarization state is computed using Jones calculus. The following Jones vector defines the incident polarized light:

$$E_i = \begin{pmatrix} 0 \\ 1 \end{pmatrix}. \tag{4.51}$$

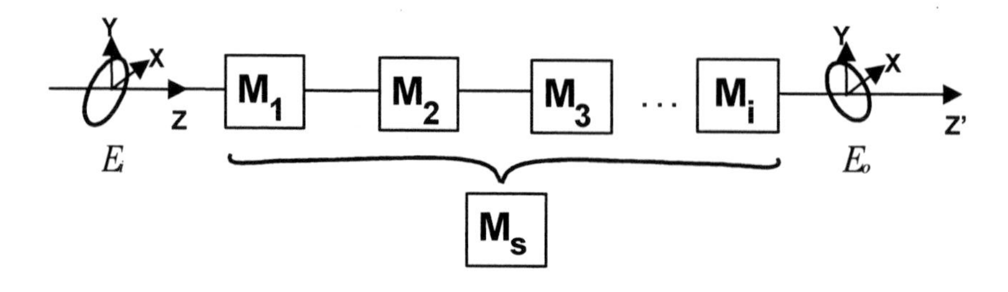

Figure 4.33 Integrated effects of the stress field may be represented by a Jones matrix.

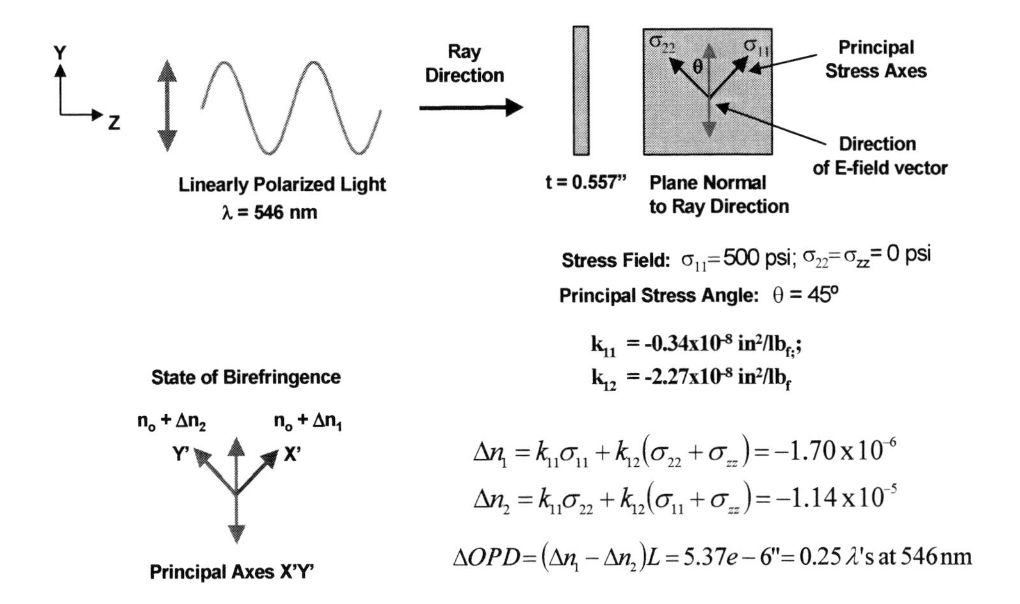

Figure 4.34 Example of stress birefringence.

The angle between the coordinate system defining the incident Jones vector and principal stress directions yields the rotation matrix:

$$\mathbf{R}(\gamma) = \begin{bmatrix} 0.707 & 0.707 \\ -0.707 & 0.707 \end{bmatrix}. \tag{4.52}$$

The change in the indices of refraction are given as

$$\Delta n_1 = k_{11}\sigma_{11} + k_{12}(\sigma_{22} + \sigma_{zz}) = (-0.34 \times 10^{-8})(500) = -1.70 \times 10^{-6}, \tag{4.53}$$

and

$$\Delta n_2 = k_{11}\sigma_{22} + k_{12}(\sigma_{11} + \sigma_{zz}) = (-2.27 \times 10^{-8})(500) = -1.14 \times 10^{-5}. \tag{4.54}$$

The resulting phase change in radians for the principal directions is given by

$$\delta_1 = \frac{2\pi \Delta n_1 t}{\lambda} \approx -0.28 \text{ rad}, \tag{4.55}$$

and

$$\delta_2 = \frac{2\pi \Delta n_2 t}{\lambda} \approx -1.84 \text{ rad.} \tag{4.56}$$

This yields the retarder matrix

$$\mathbf{R}(\delta) = \begin{bmatrix} e^{i(-0.28)} & 0 \\ 0 & e^{i(-1.85)} \end{bmatrix}. \tag{4.57}$$

Using Eqs. (4.47) and (4.49) yields the output polarization state in Jones vector format:

$$E_{out} = \begin{bmatrix} 0.616 + 0.345i \\ 0.346 - 0.618i \end{bmatrix}. \tag{4.58}$$

The magnitudes of E_x and E_y are both 0.707 and the difference in phase is $\pi/4$ rad or 90 deg. Thus, the mechanical stress acting on the window converts linear light into circular polarized light.

4.4.6. Stress birefringence and optical modeling

Computing optical performance due to three-dimensional stress birefringence requires the updated dielectric impermeability tensor be known at each point in the optical element. This may be accomplished by converting the finite element stress distribution into an index ellipsoid map using the stress-optical coefficient matrix and the nominal optical properties of the material as demonstrated in Fig. 4.35. Interpolation routines may then be used to determine the optical properties for an arbitrary ray. This capability is currently being added to commercial optical design codes.

A second method is to use the system Jones matrix, $\mathbf{M}_s$, which represents the effective optical retarder properties, to derive the birefringence, orientation, and ellipticity values for a grid of rays using the techniques discussed in section 4.4.4.

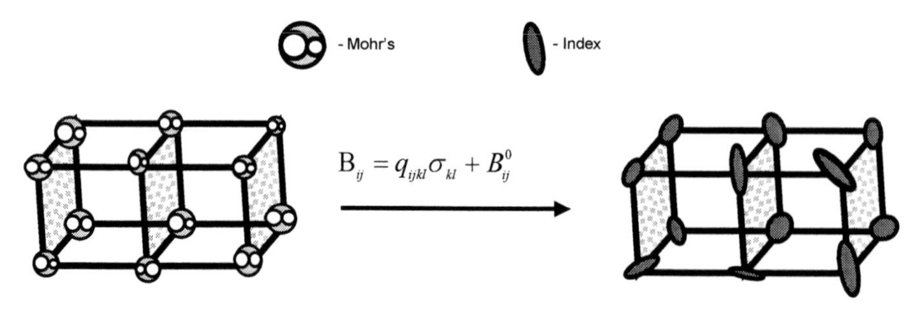

Figure 4.35 Converting a stress distribution into a 3-D birefringence model.

The birefringence and orientation values may be used to create CODE V stress birefringence interferogram files, which are an approximate technique to model the effects of a spatially varying stress files based on a linear retarder model.[14]

Examples of computing optical errors due to mechanical stress using stress-birefringence interferogram files are provided in Refs. 15 and 16. In Ref. 16, the polarization errors due to the effects of temperature on a doublet collimating lens for a telecommunication component was explored as a function of glass type. The doublet stress distribution is shown in Fig. 4.36 and the resulting birefringence and wavefront maps are shown for both front and rear elements in Figure 4.37.

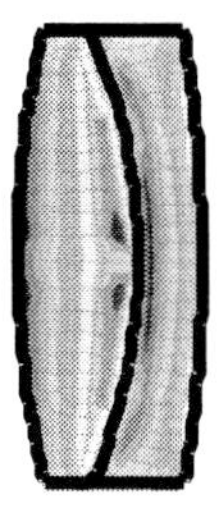

Figure 4.36 Stress field in doublet element.

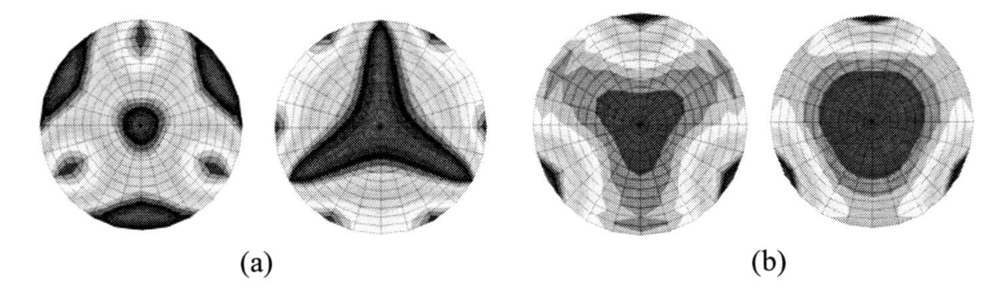

Figure 4.37 Optical errors due to mechanical stress in doublet. (a) Birefringence maps, and (b) wavefront error maps for front and rear element, respectively.

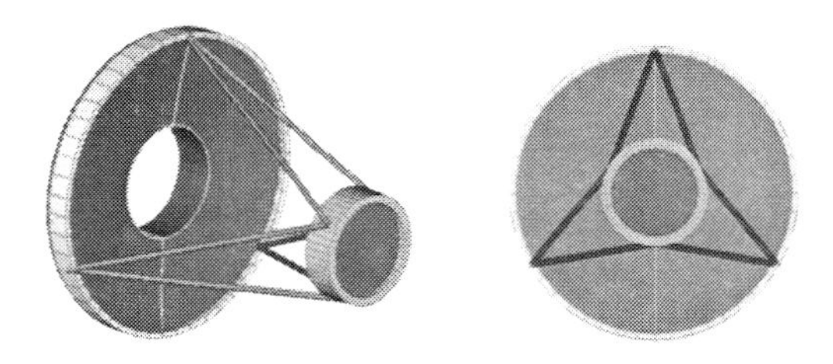

Figure 4.38 Cassegrain telescope primary mirror obscurations.

4.5 Mechanical Obscurations

In many instances, optical elements and optical support structures block a portion of the incident light passing through the optical system. For example, the secondary mirror and metering structure for a Cassegrain telescope obstructs light from reaching the primary mirror as shown in Fig. 4.38. These obscurations increase the blur diameter of an image point by scattering light normal to the boundary of the obscuration. It is important in the mechanical design effort to be able to compare the resulting image degradation due to various mechanical configurations. This section describes an approximate technique to predict the effects of obscurations on optical performance as measured by the encircled energy function.

4.5.1 Obscuration periphery, area, and encircled energy

There are two primary factors controlling the percentage of light that is diffracted by an obscuration. The periphery of the obscuration dictates the amount of energy that is diffracted. As the periphery increases, the amount of diffracted energy increases. The area of the obscuration controls the amount of energy transmitted through the optical system. As the area of the obscuration increases, the transmitted energy decreases, and a larger percentage of the light is diffracted (for the same periphery). The exact mathematical formulation to compute the effects of diffraction is typically complex. However, a simple approximation based on the ratio of the total obscuration periphery, *P*, to the total area of the transmitting aperture, *A*, given as *R*, may be used to compare mechanical design concepts. An example calculation of this ratio is illustrated in Fig. 4.39. The normalized encircled energy, *EE*, as a function of *R* is given below as

$$EE(r_o) = 1 - \frac{\lambda f}{2\pi^2 r_o} R, \qquad (4.59)$$

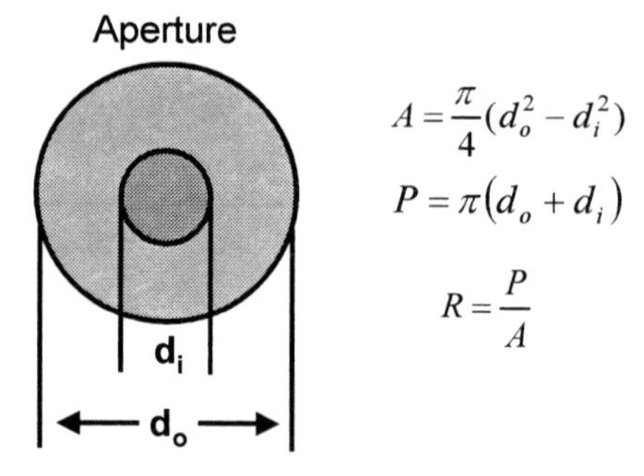

Figure 4.39 Ratio of the aperture periphery to the transmitting area controls the percentage of diffracted light.

where λ is the wavelength, f_{eff} is the effective focal length of the optical system, and r_o is the radial coordinate on the focal plane.[17] This approximation allows the encircled energy to be computed as a function of radial extent and is valid for arbitrary aperture shapes for most practical imaging applications assuming a uniformly illuminated aperture. Mechanical design trades may then be performed to evaluate mechanical support and mounting structures as a function of image quality. In addition, the ratio R may be used as a structural constraint in optimization solutions. Other spider design equations are discussed by Harvey.[18] A more detailed evaluation of the effects of obscurations on image quality may be performed using optical design software.

4.5.2 Diffraction effects for various spider configurations

The encircled energy approximation is used to compare five spider configurations for a Cassegrain telescope support structure. Each of the spider configurations has the same total cross-sectional area. The encircled energy is plotted for each configuration in Fig. 4.40.

The three-vane design provides the best optical performance with the three tangential and four-vane designs not far behind. The six and eight-vane configurations show a significant increase in diffracted energy. This result is expected given the significant increase in periphery versus area for each additional vane.

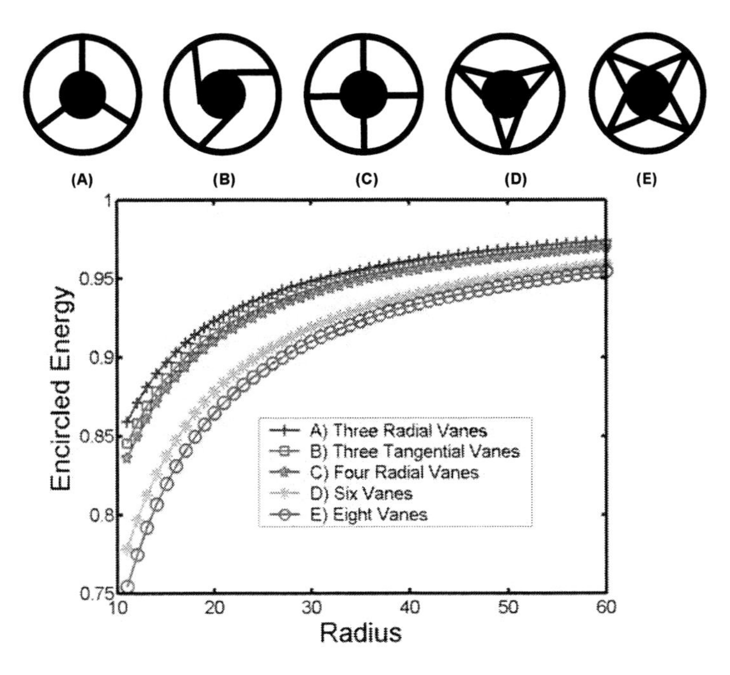

Figure 4.40 Comparison of encircled energy vs. spider configurations of constant area.

4.5.3 Diffraction spikes

Each spider configuration produces its own characteristic diffraction pattern as shown in Fig. 4.41. Notice how the diffraction pattern of the three-vanes produces six spikes in the diffraction pattern, whereas the four-vane configuration produces only four spikes. This is explained by the fact that light is diffracted in both directions normal to the vane. Hence, for the three-vane configuration with vanes at 0, 120, and 240 deg, light is scattered in six directions. The vane at 0 deg scatters light at 90 and 270 deg, the vane at 120 deg scatters light at 30 and 210 deg, and the vane at 240 deg scatters light at 150 and 330 deg. The reason you only get four directions, not eight, with four-vanes is that half of the directions are degenerate. Use of curved spider legs eliminates the diffraction spikes resulting in a rotationally symmetric diffraction image.[19] However, this does not necessarily result in improved optical performance.

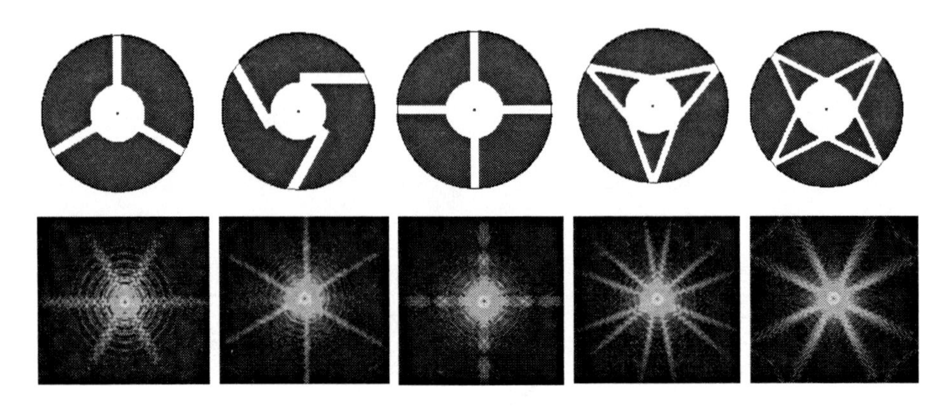

Figure 4.41 Various support structure configurations and the resulting PSF.

References

1. CODE V is a product of Optical Research Associates, Pasadena, CA
2. ZEMAX is a product of Focus Software, Inc., Tucson, AZ.
3. Genberg, V. L., Michels, G. J., "Opto-mechanical analysis of Segmented and Adaptive Optics," *Proceedings of SPIE*, **4444**, Bellingham, WA (2001).
4. Genberg, V. L., "Shape function interpolation of 2D and 3D finite element results," *Proceedings of 1993 MSC World User's Conference*, Los Angeles, CA (1993).
5. Genberg, V. L., "Ray tracing from finite element results," *Proceedings of SPIE,* **1998**, Bellingham, WA (1993).
6. Wolverton, T., Brooks, J., "Structural and optical analysis of a landsat telescope mirror," *Proceedings of MSC World User's Conf.,* MacNeal-Schwendler, Los Angeles (1987).

7. Genberg, V. L., "Structural analysis of optics," *Handbook of Optomechanical Engineering*, CRC Press (1997).
8. Miller, J.L., *Principles of Infrared Technology*, Chapman and Hall, New York, (1994).
9. Lohmann, A.W., Paris, D.P., "Influence of longitudinal vibrations on image quality," *J. Applied Optics*, **4**(4) (1965).
10. Doyle, K. B. Forman, S. E., Cerrati, V. J., Sultana, J. A., "Optimal structural design of the airborne infrared imager," *Proceedings of SPIE,* **2542**, Bellingham, WA (1995).
11. Wulich. D, Kopeika, N. "Image resolution limits resulting from mechanical vibrations," *J. Opt. Eng.*, pp. 529–533 (1987).
12. Herting, D.N., *Advanced Dynamic Analysis User's Guide*, The MacNeal-Schwendler Corporation, Los Angeles (1997).
13. *Schott Optical Glass Technical Information,* **15**, **20**, Schott Optical Glass Technologies Inc., Duryea, PA.
14. Doyle, K. B., Genberg, V. L., Michels, G. J., "Numerical methods to compute optical errors due to stress birefringence," *Proceedings of SPIE,* **4769**, Bellingham, WA (2002).
15. Doyle, K. B., and Bell, William, "Wavefront and polarization error analysis of a telecommunication optical circulator," *Proceedings of SPIE,* **4093**, Bellingham, WA (2000).
16. Doyle, K. B., Hoffman, J. M., Genberg, V. L., Michels, G. J., "Stress birefringence modeling for lens design and photonics," Invited Paper, *Proceedings of the International Optical Design Conf.*, Tucson, Arizona (2002).
17. Clark, P. D., Howard, J. W., Freniere, E. R., "Asymptotic approximation to the encircled energy function for arbitrary aperture shapes," *J. Applied Optics* **23**(2), 1984.
18. Harvey, J. E., Ftaclas, C., "Diffraction effects of secondary mirror spiders upon telescope image quality," *Proceedings of SPIE,* **965** (1988).
19. Richter, J. L., "Spider diffraction: A comparison of curved and straight legs," *J. Applied Optics*, **23**(12) (1984).

≺Chapter 5≻ Optothermal Analysis Methods

The physical properties of an optical system are modified when experiencing changes in temperature. Thermo-elastic effects modify the dimensional characteristics, and thermo-optic effects change the index of refraction of the optical materials. Predicting optical performance in high-performance optical systems typically requires the coupling of the thermal, structural, and optical analyses to account for detailed thermal interactions. This chapter discusses integrating techniques along with thermal modeling methods to allow optical performance to be predicted as a function of complex temperature distributions.

5.1. Thermo-Elastic Analysis

Temperature changes in an optical system cause dimensional and positional changes in the optical components due to thermo-elastic effects. This includes changes in optical element thickness, diameter, radii of curvature, and higher-order surface deformations. These departures from the nominal optical system prescription affect optical performance.

The material property dictating the expansions and contractions is the linear coefficient of thermal expansion (CTE), denoted by α. The CTE varies considerably depending on the type of material. For instance, the CTE of common optical mounting materials include Invar 36 at ~ 1.0 ppm/C, titanium at 8.8 ppm/C, and aluminum at 23.6 ppm/C. The CTE of optical glasses include fused silica at 0.5 ppm/C, BK7 at 7.1 ppm/C, and FK54 at 14.6 ppm/C. The CTE of plastics and epoxies may be an order of magnitude greater than metals or glasses with CTEs in the hundreds of ppm/C.

5.1.1 CTE temperature dependence

The temperature dependence of the coefficient of thermal expansion must be accounted for in extremely sensitive optical systems or for optical systems experiencing large temperature swings. For example, the variation in CTE with temperature for aluminum, beryllium, and fused silica is shown in Fig. 5.1. A nonlinear finite element solution is required if the thermo-elastic response quantities are desired at incremental times as the optical instrument cools down or heats up.

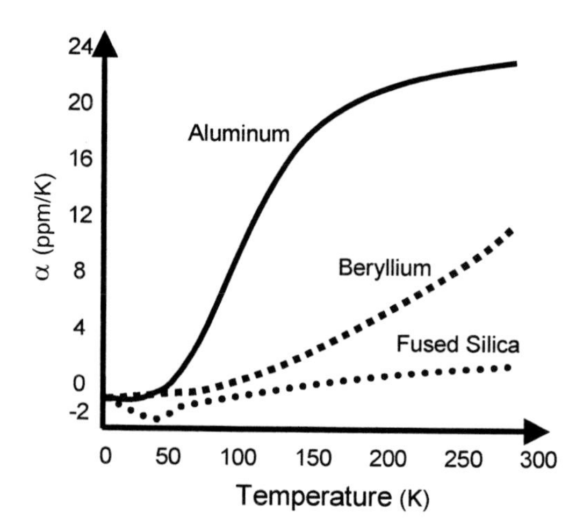

Figure 5.1 CTE as a function of temperature for selected materials.

A linear finite element solution may be run to compute the response quantities at one predefined temperature using an effective CTE or an effective thermal load vector. The total thermal strain at the temperature of interest is required, but not the thermal strain path. The thermal strain, ε_t, for a material changing temperature from T_1 to T_2 is given by the integral expression

$$\varepsilon_t = \int_{T_1}^{T_2} \alpha(t)dt . \tag{5.1}$$

This expression is equivalent to computing the area under the CTE versus temperature curve as shown in Fig. 5.2. An effective CTE or thermal load vector may then be computed using the thermal strain. The effective CTE, α_{eff}, is given as

$$\alpha_{eff} = \frac{\varepsilon_t}{T_2 - T_1} . \tag{5.2}$$

An effective thermal load vector, ΔT_{eff}, may be computed as

$$\Delta T_{eff} = \frac{\varepsilon_t}{\alpha_o} , \tag{5.3}$$

where α_o is the coefficient of thermal expansion at temperature, T_1. Using an effective CTE or thermal load vector assumes that all other mechanical properties are linear over the temperature range. This approach may be repeated for several temperatures of interest in place of a nonlinear analysis.

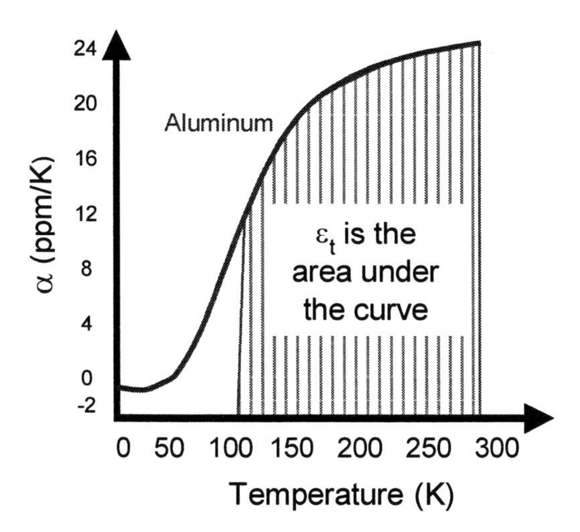

Figure 5.2 Thermal strain is the area under the CTE vs. temperature curve and is shown above for aluminum cooling from room temperature to 100 K.

5.1.2 CTE inhomogeneity

The spatial variation of the CTE in optical elements and mounting structures is often a concern for sensitive optical systems. The CTE inhomogeneity may be due to the fabrication of the raw materials and/or the micro-mechanical variations in grain size and orientation. For uniform temperature changes, the effects of CTE inhomogeneity are equivalent to thermal gradients acting on the optical system. The CTE may be varied in a finite element model by modifying the value on the material card definition. However, only one CTE variation can be analyzed in a single run, as a new stiffness matrix must be computed due to the change in material properties. A more efficient technique to study CTE variations is to represent the CTE variation as thermal load vectors as illustrated in Fig. 5.3. This allows multiple thermal load vectors to be input in a single finite element solution using subcases. This offers time and computational savings by inverting the stiffness matrix (which is the major expense) only once to compare various CTE spatial distributions.

The equivalent thermal load vector, ΔT_{equ}, may be computed by the following relationship:

$$\Delta T_{\text{equ}}(x,y,z) = \frac{\alpha(x,y,z)}{\alpha_o}\Delta T\,, \tag{5.4}$$

where $\alpha(x,y,z)$ is the CTE at coordinate location x, y, and z, and α_o is the nominal CTE value. This technique is particularly useful for studying random variations in the CTE. A program or spreadsheet may be used to generate the random thermal load vectors.

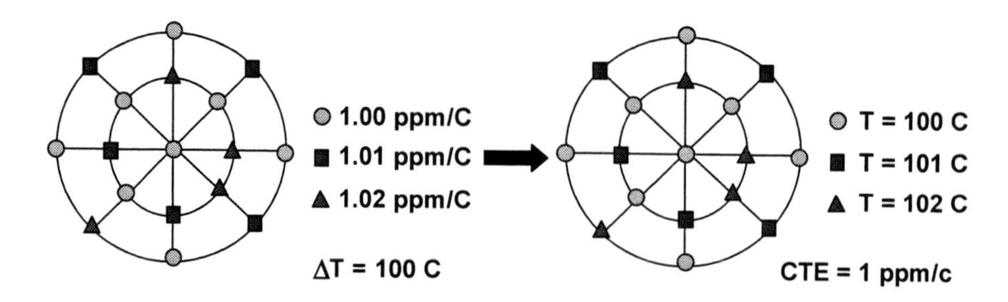

Figure 5.3 Equivalent models accounting for spatial variations in CTE.

A combined thermal load vector may be computed to account for the CTE variation with temperature and CTE inhomogeneity using

$$\Delta T\left(x,y,z,T_2\right)=\frac{1}{\alpha_o}\int_{T_i}^{T_2}\alpha(x,y,z,t)dt\,, \tag{5.5}$$

where $\alpha(x,y,z,T_2) = \alpha(x,y\ z,\ T_1) + \Delta\alpha(T_2)$, which assumes the thermal variation is not spatially dependent and $\Delta\alpha(T_2)$ is equal to the temperature-dependent deviation in CTE from $\alpha(x,y,z,T_1)$.

For the special case of a uniform plate, an effective thermal soak, ΔT_{eff}, and a through-the-thickness thermal gradient, T'_{eff}, may be defined as follows:

$$\Delta T_{\text{eff}}\left(x,y,T_2\right)=\frac{1}{\alpha_o t}\int_{t/2}^{t/2}\int_{T_1}^{T_2}\alpha(x,y,z,t)dtdz\,, \tag{5.6}$$

and

$$T'_{\text{eff}}\left(x,y,T_2\right)=\frac{12}{\alpha_o t^3}\int_{-t/2}^{t/2}\int_{T_1}^{T_2}\left[\alpha\left(x,y,z,t\right)\right]dtzdz\,. \tag{5.7}$$

5.2 Thermo-Optic Analysis

The thermo-optic coefficient, *dn/dT*, defines the change in the index of refraction of an optical material as a function of temperature. The relative *dn/DT* value, denoted by dn_{rel}/dT, is the change in the index of refraction of the optical material relative to air. (This value is also a function of pressure, which may be important for optical systems experiencing altitude changes.) The absolute *dn/DT*, denoted by dn_{abs}/dT, is the change in the index of refraction relative to vacuum. For optical systems in air, the relative thermo-optic coefficient is convenient to use since the change in the index of the air does not have to be specified. When using dn_{abs}/dT, the change in the index of the optical medium must also be taken into account. The relationship between the two thermo-optic coefficients for an optical material of index, n, is

$$\frac{dn_{\text{abs}}}{dT} = n\frac{dn_{\text{air}}}{dT} + n_{\text{air}}\frac{dn_{\text{rel}}}{dT}. \tag{5.8}$$

Thermo-optic coefficients vary widely among glass types with positive and negative values. For example, the absolute thermo-optic coefficients for several Schott glasses at room temperature at a wavelength of 546 nm are listed in Table 5.1.

Table 5.1 Absolute thermo-optic coefficients at 546-nm wavelength and room temperature.

GLASS TYPE	DN/DT (PPM/C)
BK7	1.6
LaF2	–0.7
SF1	6.4
FK51	–7.0
LaK23	–2.0
BaF3	2.1
LaSF3	5.2
PSK2	1.3
FK54	–6.9

5.2.1 Wavefront error

Changes in the index of refraction of an optical material cause wavefront errors in the optical system. The change in optical path for a given ray traveling through an optical element in air with a uniform temperature change, ΔT, is computed as

$$\text{OPD} = \frac{dn_{\text{rel}}}{dT}(\Delta T)L, \tag{5.9}$$

where L is the distance the ray travels in the optical element. For a nonuniform temperature change, the OPD may be computed by incrementally summing the optical path difference given by

$$\text{OPD} = \sum_{m=1}^{m=\text{total}} \frac{dn_{\text{rel}}}{DT} L_m \Delta T_m, \tag{5.10}$$

where m is the number of increments, L_m is the incremental distance traveled by the ray, and T_m is the temperature change in each increment. This may be repeated for a grid of rays to provide an OPD map across the pupil.[1]

5.2.2 Sellmeier dispersion equation

The thermo-optic coefficient is a function of both wavelength and temperature as given by the Sellmeier Dispersion equation[2] below:

$$\frac{dn_{\text{abs}}(\lambda,T)}{dT}=\frac{n^2(\lambda,T_o)-1}{2n(\lambda,T_o)}\left(D_0+2D_1\Delta T+3D_2\Delta T^2+\frac{E_0+2E_1\Delta T}{\lambda^2-\lambda_{TK}{}^2}\right), \tag{5.11}$$

where T_o is the reference temperature at 20°C, ΔT is the change in temperature, λ is the wavelength in a vacuum in microns, λ_{TK} is the average effective resonance wavelength in microns, D_0, D_1, E_0, and E_1 are material dependent constants. For example, the absolute thermo-optic coefficient is plotted as a function of wavelength and temperature for the Schott glass FK5 in Figs. 5.4 and 5.5, respectively.

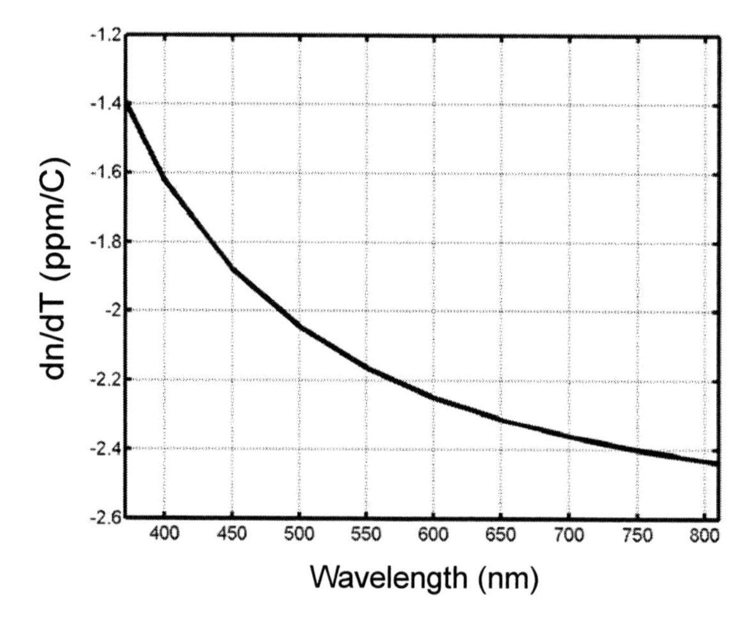

Figure 5.4 Absolute thermo-optic coefficient for Schott Glass FK5 as a function of wavelength.

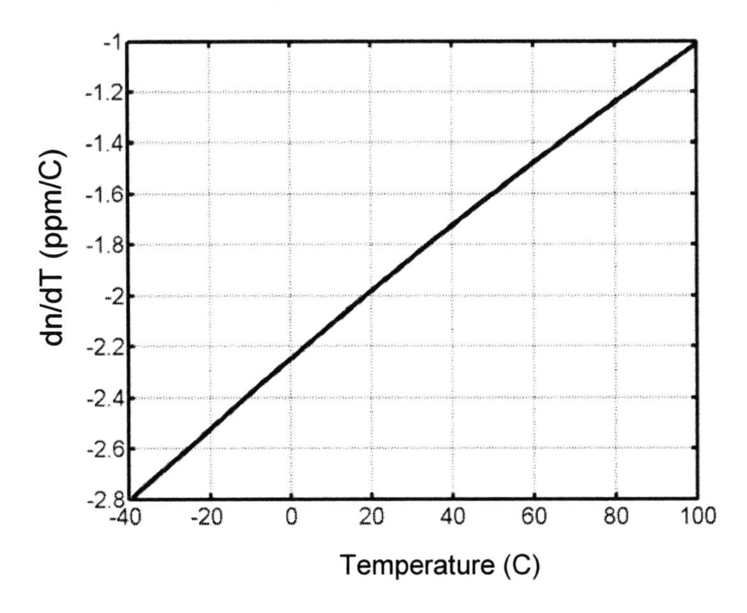

Figure 5.5 Absolute thermo-optic coefficient for Schott Glass FK5 as a function of temperature.

5.3 Effects of Temperature on Optical System Performance

The most common optical error due to uniform temperature changes is to put the system out of focus. This is particularly true for complex systems and optical systems with long focal lengths. Radial gradients also introduce focus error along with higher-order aberrations. For more complex temperature gradients, multidisciplinary analyses are required to predict optical performance. Discussed below are the relationships between the CTE, thermo-optic coefficient, and optical errors for simple temperature loads.

5.3.1 Thermal soak conditions

For a single thin lens element in air, the change in focal length, Δ_f, for a uniform temperature change is given by

$$\Delta_f = \left[\alpha - \left(\frac{1}{n-1}\frac{dn_{\text{rel}}}{dT}\right)\right] f\Delta T, \tag{5.12}$$

where f is the focal length of the lens. The term in the brackets is known as the opto-thermal expansion coefficient or the thermal-glass constant, η. This coefficient accounts for both shape and index changes.

The change in focal length may be compensated or balanced by the change in position of the image plane due to the thermal expansion and contraction of the metering structure, α_m, as shown below:

$$\Delta_f = (\eta - \alpha_m) f \Delta T . \quad (5.13)$$

Typically, it is much more difficult to passively compensate infrared systems as compared to visible systems since the thermal-glass constant for infrared materials is generally much larger than the CTE of conventional housing materials.

A system thermal-glass constant, η_s, may be specified for lens elements in contact as the following:

$$\eta_s = \sum_{i=1}^{n} \frac{f_s}{f_i} \eta_i , \quad (5.14)$$

where f_s is the system focal length, and f_i are the focal lengths of the individual elements.

5.3.1.1 Focus shift of a doublet lens

The focus error for a doublet lens, shown in Figure 5.6, is computed for several potential metering structure materials. The optical system is imaging an object at infinity with a wavelength of 546 nm, has a focal length of 100 mm, and an *f*/# of 3.0.

The CTE, relative *dn*/*DT*, thermal-glass constant, and the focal length of the individual lens elements are listed in Table 5.2.

Using Eq. (5.14), a system-level thermal-glass constant is computed for the doublet as is given by

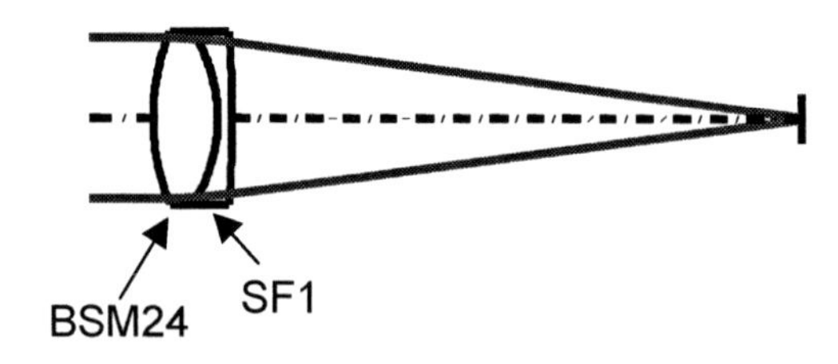

Figure 5.6 Doublet lens element.

Table 5.2 Doublet properties.

DOUBLET	α	DN_{REL}/DT	η	F(MM)
BSM24	6.5	4.3	–0.4	42.0
SF1	8.1	7.9	–2.8	–68.1

$$\eta_s = \sum_{i=1}^{n} \frac{f_s}{f_i} \eta_i = 3.2 \text{ ppm}/c. \tag{5.15}$$

A reasonable goal is to select a mounting material that minimizes the focus error such that diffraction-limited performance is maintained. This is given by the Rayleigh quarter-wave criterion and is known as the diffraction-limited depth-of-focus and is expressed as

$$\delta = \pm 2\lambda \left(f/\# \right)^2 = \pm 9.8 \ \mu\text{m}, \tag{5.16}$$

where δ is the longitudinal distance in which the image may be offset from the ideal image location. Hence, to maintain diffraction-limited performance, the following condition must be met:

$$\Delta_{\text{focus}} = \left(\eta_s - \alpha_s \right) f \Delta T \leq 9.8 \ \mu\text{m}. \tag{5.17}$$

Three materials were considered for the housing: aluminum, stainless steel 416, and Invar 36. The CTE of each material, the change in focus per degree Celsius, and the total temperature range, ΔT, in which the system maintains diffraction-limited performance is given in Table 5.3.

Table 5.3 Diffraction-limited performance.

MATERIAL	ΔF (μM/C)	ΔT
Aluminum	**–2.04**	**4.8**
SS416	**–0.67**	**14.6**
Invar36	**0.22**	**44.5**

5.3.2 Radial gradients

For temperature gradients that are constant through the thickness of the optical element such as radial gradients, the wavefront error or OPD at a given point may be approximated by the following expression:

$$\text{OPD} = \left[\alpha (n-1) + \frac{dn_{\text{rel}}}{dT} \right] t \Delta T. \tag{5.18}$$

The expression in the bracket of Eq. (5.18), is referred to as the thermo-optic constant (do not confuse with the thermo-optic coefficient) and is an approximate measure of the sensitivity of the optical element to radial gradients.[3] Note that in Eq. (5.18), the CTE and thermo-optic coefficients are added, whereas in Eq. (5.12) they are subtracted. Thus, in general, for lens assemblies whose glass types tend to minimize focus shifts due to thermal soak conditions, they are typically more sensitive to radial gradients.

5.4 Optical Design Software

For complex optical systems comprising multiple lens elements, the thermal response of an optical system may be analyzed using optical design software. Thermal analysis capabilities within optical design modeling tools account for first-order thermo-elastic and thermo-optic effects that allow optical performance to be evaluated and optimized over temperature. This includes being able to simulate thermal soak conditions and account for the expansions and contractions of the spacers and mount. Use of dummy surfaces allows mechanical mounting arrangements that are more complex to be modeled. CODE V provides the ability to simulate radial gradients. Axial gradients may be approximated using multiple surfaces to define a lens element. Thermo-optic effects due to complex temperature distributions may be modeled using a user-defined refractive index profile. Here the index of refraction and index gradients must be defined as functional relationships. Developments are underway in commercial optical design codes that allow the index of refraction to be defined at discrete locations in an optical element. An example of athermalizing a lens assembly using optical design software and interferogram files is discussed in Ref. 4.

5.4.1 Representing OPD maps in the optical model

Optical design codes offer various formats to represent externally derived OPD maps in the optical model. In general, this data is assigned to optical surfaces and pupils. Wavefront interferogram files are used in CODE V. The data may be in a uniform rectangular array or represented by Zernike polynomials. Zemax represents OPD maps using a grid phase surface in which the user supplies a uniform array of data and a bicubic spline is used to define the phase map.

The OPD data is typically assigned to dummy surfaces in the optical model and deviates and adds OPD to the rays that intersect the surface but does not affect the surface shape. Stress-induced wavefront error may also be represented in this manner.

When assigning OPD data to an optical model it is critical that the proper sign conventions are understood. In CODE V, for example, a positive wavefront error represents a leading wavefront as shown in Fig. 5.7. In addition, it is equally important to align and place the data at the correct location and with the proper orientation.

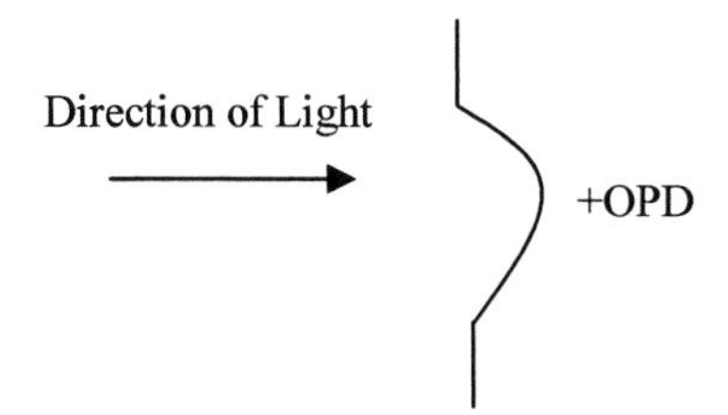

Figure 5.7 Sign convention for CODE V wavefront interferogram files.

5.5 Thermo-Optic Finite Element Models

Thermo-optic finite element models are a useful modeling technique to compute OPD maps due to complex temperature profiles. A thermo-optic model is created by modifying the material properties and boundary conditions of a 3D finite element model of the optical element.[5] The material properties are modified by replacing the coefficient of thermal expansion with the thermo-optic coefficient (relative or absolute, depending on the application), setting the elastic modulus to a value of one, and setting the shear modulus and Poisson's ratio to zero. This decouples the in-plane and out of plane effects. The nodes on the front surface are constrained in the three translational degrees of freedom, while the remaining nodes in the model are constrained in the axes normal to the optical axis (here, the optical axis is assumed to be along the z-axis). The applied load vector is the temperature field. A schematic of a thermo-optic finite element model is shown in Fig. 5.8. At the rear surface of the optical element, a finite element solution will yield a displacement map representing the optical path difference for an exiting wavefront. This displacement profile may then be post-processed similarly to optical surface deformations. For example, the OPD map may be fit to Zernike polynomials or interpolated to a uniform grid. This technique assumes the rays traverse the optical element along the optical axis or z-direction. Hence, for powered optics this technique is an approximation. An example of this method is illustrated for a lens element in Fig. 5.9.

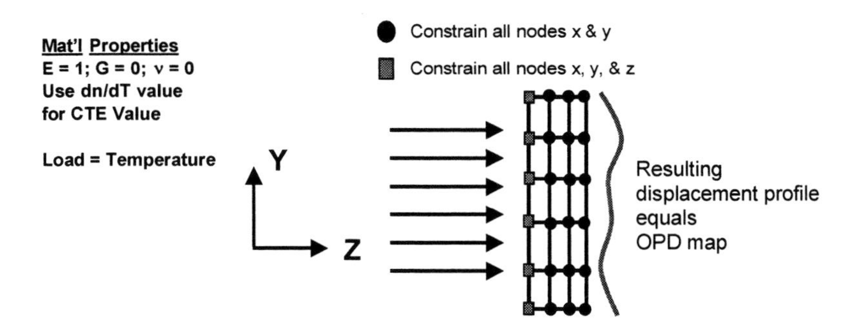

Figure 5.8 Creating a thermo-optic finite element model.

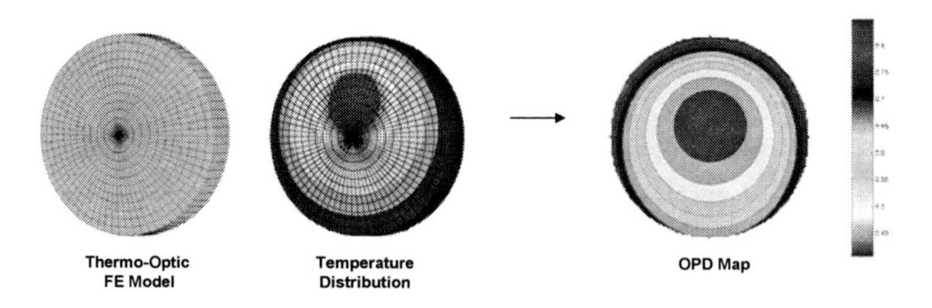

Figure 5.9 Thermo-optic finite element model resulting in an OPD map.

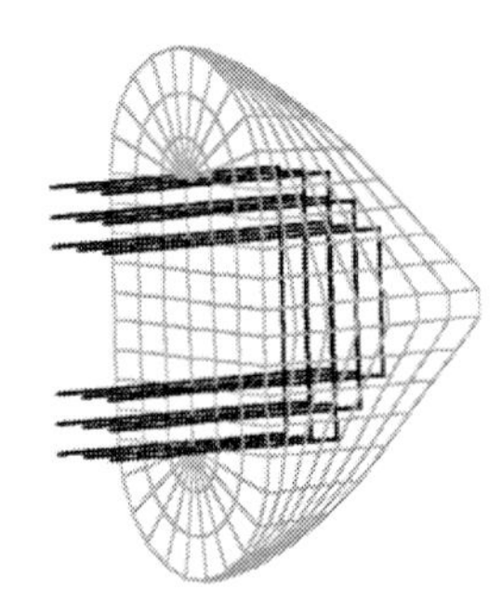

Figure 5.10 Porro prism finite element model.

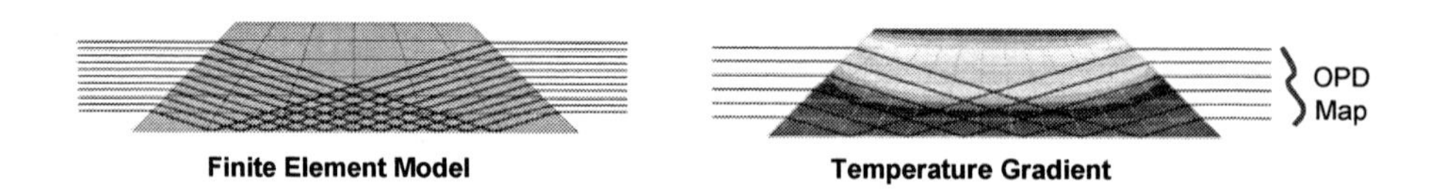

Figure 5.11 Dove prism finite element model and temperature distribution.

5.5.1 Multiple reflecting surfaces

Thermo-optic finite element models may also be used to compute wavefront maps for optical elements with multiple reflecting surfaces such as beam splitters and prisms. A finite element model of a Porro prism with two reflecting surfaces is shown in Fig. 5.10. Thermo-optic models for optical elements with multiple reflecting surfaces is typically more difficult than for single-pass transmissive elements. The ray paths must be modeled using rod elements. The OPD map of the exiting wavefront is computed by summing the axial displacements of the rod element representing a given ray. This requires mapping temperatures to the nodes of the rod elements from a 3D temperature distribution of the optical element. Mapping techniques discussed in Sec. 5.7 may be applied here. This modeling technique is demonstrated for a Dove prism in Fig. 5.11.

5.6 Bulk Volumetric Absorption

As light travels through a transmissive optical element in a typical imaging system, the majority of the light is transmitted. However, a small fraction of the light is typically absorbed in what is referred to as bulk volumetric absorption. The absorbed energy produces thermal heating. The mechanism responsible for the absorption is the oscillation of the electrons within the atomic structure, which consists of negatively charged electrons floating in bands around the positively charged nucleus. An electrostatic force, or spring, binds the electrons to the nucleus. The forcing function acting on the atomic oscillator is the time-varying electric field of an incident wavefront, which causes the electrons to vibrate analogous to a mechanical oscillator. The increase in motion of the electrons is dampened by the neighboring atoms and molecules of the material resulting in the dissipation and the absorption of energy. The greatest interaction occurs at the condition of resonance when the electrical field of the incident wavefront has the same frequency as a characteristic frequency of the optical material. The interaction described above is responsible for heating via absorption, and also is responsible for the variation in the index of refraction of a material as a function of wavelength known as dispersion.

The amount of energy absorbed via bulk volumetric absorption may be approximated by the use of Beer's law. The absorbed energy, A, is a function of the absorption coefficient, κ_λ, and the distance the wavefront travels in the optical element, d:

$$A = 1 - e^{-\kappa_\lambda d}\,. \tag{5.19}$$

An absorption coefficient for a given material and wavelength may be computed using the materials' transmission characteristics. For example, BK7 transmits 99.6% of 546 nm light. Using Beer's law, an absorption coefficient of 0.004 mm^{-1} is computed. This approach may be used to account for a spectral range by summing the watts absorbed by each individual wavelength.

Accounting for bulk volumetric absorption in solid transmissive elements using thermal analysis software may be performed using a two-stage thermal modeling process.[6] The first step is to develop a heat-rate model using plate elements of zero thickness to determine the energy absorbed. The second step involves transferring the nodal heat rates determined in the plate elements to a thermal-network model. Then, a steady-state analysis is performed to compute the temperature distribution in the optical element.

Several modeling techniques may be employed to shape, mask, and direct the radiation within the optical model. This includes using multiple heat sources of varying intensity, optical elements to focus and direct the energy beam, masks to contour the beam, and absorption coefficients that vary with angle of incidence.

5.7 Mapping of Temperature Fields from the Thermal Model to the Structural Model

Performing a thermo-elastic analysis often requires the mapping of temperatures from the thermal model to the structural model. This is an issue when the thermal and structural models do not share a common mesh that results primarily from the different physical phenomena each model represents. Several mapping techniques of varying complexity and accuracy are discussed. The user should decide which method is appropriate based on the application and sensitivity of the instrument.

5.7.1 Nearest-node methods

One approach to map temperatures from the thermal to the structural model is to transfer the temperature of the closest thermal node to the structural node. A derivative of this method is to compute a nodal average of the nearest thermal nodes to a given structural node and weight their contribution by distance. Nearest-node methods are appropriate only for the interior nodes of continuous media, since the method cannot account for boundaries, gaps, and element properties in the neighboring nodes, and these techniques cannot extrapolate to nodes that sit on edges of boundaries.

5.7.2 Conduction analysis

A convenient method to map temperatures to the structural model is to perform a finite element conduction analysis. This assumes there are more structural nodes than thermal nodes, and that a structural node exists near or, preferably, at each thermal node location. The temperatures computed by the thermal model serve as boundary conditions in the conduction analysis, as illustrated in Fig. 5.12. Advantages of this technique include accounting for gaps and element properties. However, this technique may generate boundary-condition errors. A coarse thermal model is shown with a temperature distribution in Figure 5.13(a).

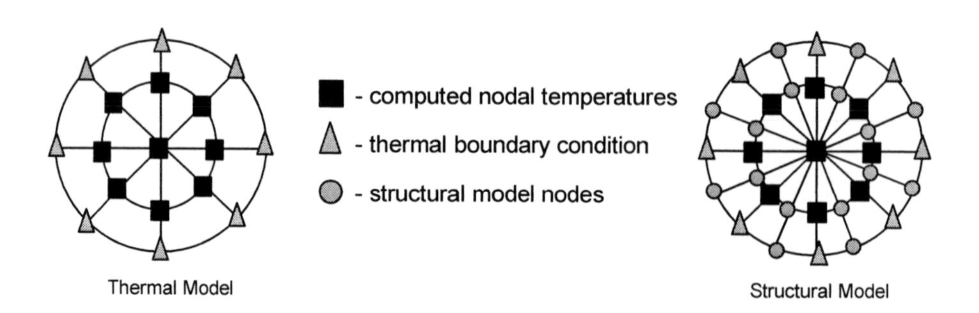

Figure 5.12 Temperature mapping using a conduction model.

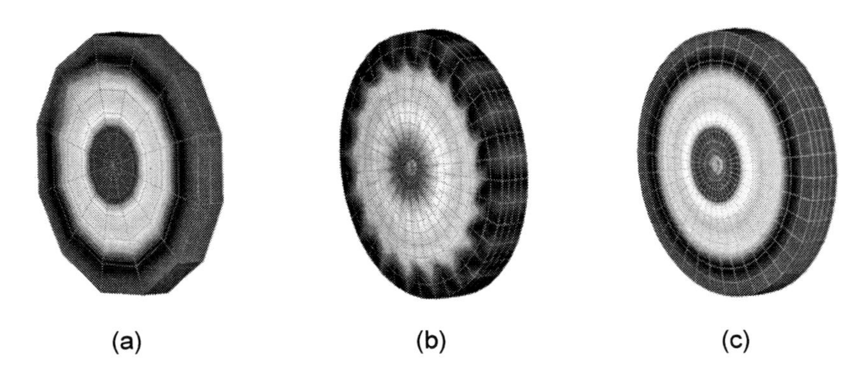

Figure 5.13 (a) Thermal model temperature distribution; (b) temperatures mapped to structural model using conduction model; (c) temperatures mapped to structural model using finite element shape functions.

The temperatures are mapped to a structural model with a finer mesh using the conduction method as shown in Figure 5.13(b). The additional thermal paths within the structural model create a spiked temperature distribution ultimately resulting in thermo-elastic and thermo-optic errors.

5.7.3 Shape function interpolation

A technique that overcomes the limitations of the previous methods is to use the finite element shape functions to interpolate the temperatures to the structural nodes.[7] This technique extracts the most information from the finite element model by using the same theory to post-process the data that was used to solve the problem. Shape function interpolation is discussed in more detail in Chapter 4. An example of interpolating temperatures from the thermal to structural model using shape functions is demonstrated in Fig. 5.13(c).

5.8 Analogous Techniques

Numerical techniques applicable to solving conduction problems may be used to solve mass-diffusion-type problems.[8] Fick's law is the governing differential equation for problems of mass diffusion, which is of the same form as Fourier's law, the governing differential equation for heat conduction. Fick's law is expressed as

$$\frac{\dot{m}}{A} = -D\frac{\partial C}{\partial x}, \tag{5.20}$$

where D is the diffusion coefficient, C is the moisture concentration, and $\dot{m}$ is the mass flux per unit time.

Fourier's law of heat conduction is given as

$$\frac{q}{A} = -k\frac{\partial T}{\partial x}, \tag{5.21}$$

where k is the thermal conductivity, T is the temperature, and q is the heat flux per unit time. Thus, using heat conduction modeling tools, mass-diffusion-type problems may be solved by applying the appropriate substitutions.

5.8.1 Moisture absorption

A common application of the above techniques is to compute the dimensional changes in plastic optics due to moisture absorption, which is a form of mass diffusion and governed by Fick's law. Using the following thermal-to-moisture analogy, thermal modeling tools may be used to compute the moisture concentration: moisture concentration is substituted for temperature, diffusivity replaces conductivity, the moisture gradient replaces the temperature gradient, and the moisture flow equals the thermal flux. In the above analysis, the thermal capacitance should be set to unity. For a known moisture concentration, a thermo-elastic analysis may be performed to compute dimensional changes. Here, the coefficient of moisture expansion (CME) is used in place of the CTE and the moisture concentration is used as the thermal load.

5.8.2 Adhesive curing

For optical elements mounted with adhesives, shrinkage during curing may cause misalignment and introduce stress in the optical element. Adhesive curing is a form of mass diffusion and the techniques applied to moisture absorption/desorption may be applied to compute the effects of shrinkage. Modeling curing requires the solvent concentration to be used in place of the temperature distribution and the coefficient of solvent shrinkage substituted for the CTE value. Generally, the coefficient of solvent shrinkage coefficients must be obtained from test data for each adhesive in consideration.

References

1. Genberg, V. L., Michels, G. J., Doyle, K. B., "Making mechanical FEA results useful in optical design," *Proceedings of SPIE*, **4769**, Bellingham, WA (2002).
2. *Schott Glass Catalog,* Schott Glass Technologies, Inc., Duryea, PA (1995).
3. Rogers, P.J., Roberts, M., "Thermal compensation techniques," *Handbook of Optics, 2nd Ed., Volume I,* McGraw-Hill, Inc., New York (1995).
4. Doyle, K. B., Hoffman, J. M., "Athermal design and analysis for WDM applications," *Proceedings of SPIE*, **4444**, Bellingham, WA (2001).
5. Genberg, V. L., "Optical path length calculations via finite elements," *Proc. of SPIE*, **748**(14), Bellingham, WA (1987).

6. Doyle, K. B., Bell, W. M.,"Thermo-elastic wavefront and polarization error analysis of a telecommunication optical circulator," *Proceedings of SPIE* **4093**, Bellingham, WA (2000).
7. Genberg, V. L., "Shape function interpolation of 2D and 3D finite element results," *Proc. of 1993 MSC World User's Conference*, Los Angeles, 1993.
8. Genberg, V. L., "Solving field problems by structural analogy," *Proceedings of Western New York Finite element User's Conference*, STI, Rochester, NY (1986).

<Chapter 6>
Adaptive Optics Analysis Methods

This chapter presents methods and concepts relevant to finite element simulation of adaptive optical systems.

6.1 Introduction

In an adaptive optical system, image motion and aberrations are reduced by moving and deforming one or more optical surfaces. Such adjustment may be made continuously to compensate for drifting or harmonic disturbances. Often times such adjustments are made in response to a measurement of the optical performance of the system. Figure 6.1 shows a schematic of an adaptive telescope in which aberrations caused by a turbulent atmosphere are corrected. Before reaching the image plane, some of the light is split into a wavefront sensor. Measurements from the wavefront sensor are sent to a controller that statistically predicts how the deformable mirror should be actuated to best correct the incoming aberrations. In such a system, the ability of the algorithm used to predict accurate control commands to best correct the induced aberrations is critical to the net optical performance. It is equally important, however, for the deformable mirror to be able to deform into shapes that will be required to correct the aberrations induced by the atmosphere. Therefore, predictions of the deformable mirror's performance are of great interest to engineers designing such a system.

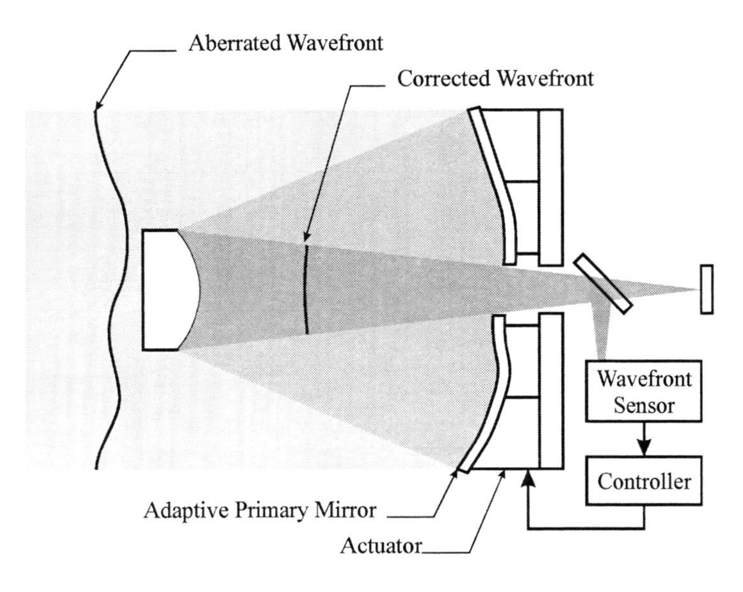

Figure 6.1 Schematic illustration of an adaptive optical system.

It is important to note that many scenarios that are not thought of as adaptive control problems can use the same simulation techniques as are used for adaptive control applications. The support of a large optic mounted on multiple airbags with adjustable pressures is one such example.

6.2 Method of Simulation

The method of simulation of adaptive control that is presented here is limited to static single-pass adaptive control. Issues associated with closed loop control such as feedback and dynamic response are outside the scope of this text. It is also assumed that the corrected performance is linearly related to each actuator's influence on the system.

6.2.1 Determination of actuator inputs

The goal in the determination of the actuator inputs is to compute a set of actuator forces or displacements that minimizes the surface error of a deformed optical surface.[1] Once actuator inputs have been found, the residual-surface error, correctability, and other performance metrics may be found. Figure 6.2 shows that the corrected surface is equal to the sum of the uncorrected deformed surface plus each of the actuator influence functions scaled by each actuator's respective actuator input. We start by writing the expression for the *ith* node's corrected displacement, ds_i^{Corr}, for the optical-surface FEM that has been corrected by the vector of actuator-control inputs **x**:

$$ds_i^{Corr} = ds_i + \sum_j \mathbf{x}_j dx_{ji} , \tag{6.1}$$

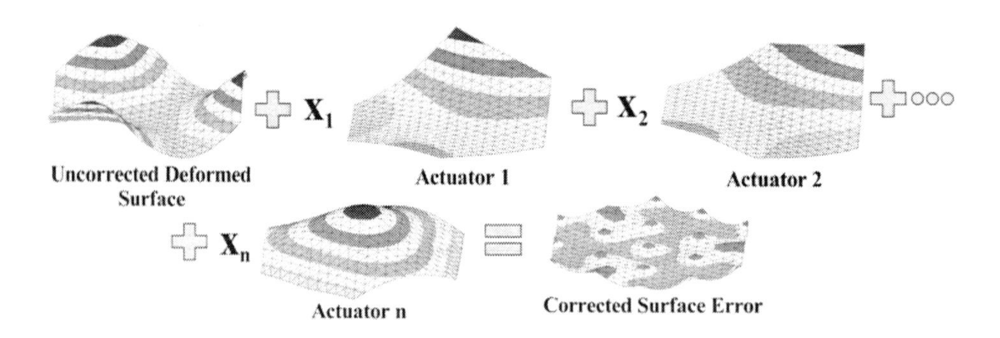

Figure 6.2 Illustration of adaptive control simulation. The unactuated deformed surface plus all of the influence functions scaled by the actuator inputs equals the corrected surface.

where ds_i is the uncorrected displacement of the ith node, x_j is the variable actuator input for the jth actuator, and dx_{ji} is the displacement of the ith node for the jth actuator's influence function. The influence function for a particular actuator is the deformed surface due to a unit input of that actuator while all other inputs are zero. Now, we write an expression of the surface mean square of the corrected optical surface, E:

$$E = \sum_i w_i \left(ds_i + \sum_j x_j dx_{ji} \right)^2 , \quad (6.2)$$

where w_i is the area weighting of the ith node. The weighting factors must be scaled such that their sum is equal to unity.

$$w_i = \frac{A_i}{\sum_i A_i} . \quad (6.3)$$

To solve for the actuator inputs that minimize the mean square error, E, we take derivatives of Eq. (6.2) with respect to each actuator input, x_j, and set each resulting equation equal to zero. This results in the following linear system:

$$[H]\{X\} = \{F\} , \quad (6.4)$$

where

$$H_{jk} = \sum_i w_i dx_{ji} dx_{ki} ,$$

and

$$F_k = \sum_i w_i ds_i dx_{ki} . \quad (6.5)$$

Once the actuator inputs have been found from Eq. (6.4), Eq. (6.2) may be used to compute the mean-square surface error. The surface rms error is the square root of the mean-square error. The corrected nodal-displacements given in Eq. (6.1) may be used for surface peak-to-valley computation, fitting to Zernike polynomials, or interpolation to an array. The latter two processes may be used to generate input to an optical analysis program.

Notice that the actuator-influence matrix, dx_{ji}, may contain influence function displacement vectors of any scalable surface modification process. Such influence-function displacement vectors include those due to force or displacement actuators, of course, but could also represent other processes. For example, if an astronomical telescope includes capability to adjust the position of a downstream fold mirror to recalibrate the system for best focus, such adjustment may be equivalent to adding the $2\rho^2-1$ Zernike term to the primary mirror's optical prescription. Therefore, an actuator-influence function that represents such a profile may be included in the actuator-influence matrix so that the effects of adaptive and focus controls are considered simultaneously. Such displacement vectors in the actuator influence-matrix are referred to as augment actuators.

Notice that if the actuator-influence matrix, dx_{ji}, lacks influence function displacement vectors that represent rigid-body motions of the optical surface, then the system will not be able to perfectly correct for rigid-body motions in the disturbance. Such rigid-body-motion content in the actuator-influence matrix may be represented by three individual displacement vectors such as three displacement actuator-influence functions of a force-actuated system mounted on three-displacement actuators. However, rigid-body-motion content may also be represented by a combination of displacement vectors as in a system using many displacement actuators. For optics that rely on rigid-body-motion disturbances being removed elsewhere in the optical system, augment actuators representing the rigid-body motions of the optical surface should be used to represent a correction of the rigid-body-motion disturbances. The lack of rigid-body-motion content in the actuator-influence matrix will severely affect performance predictions in most systems.

6.2.2 Characterization metrics of adaptive optics

A common metric with which adaptively corrected optical performances are characterized is correctability. Correctability is a relative measure of the decrease in surface deformation from an uncorrected state to a corrected state. However, correctability may be defined in a variety of different ways. Each definition quantifies a unique aspect of how correctable an optic is for a given uncorrected deformation and actuator configuration. To illustrate the various definitions of correctability we first define four rms surface characterizations in Table 6.1.

Table 6.1 Surface rms definitions.

SURFACE RMS SYMBOL	DEFINITION
R_u	**Surface rms of the uncorrected surface.**
R_b	**Surface rms of the uncorrected input surface minus the best fit plane motions.**
R_f	**Surface rms of the uncorrected input surface minus the best fit plane motions and best fit power.**
R_c	**Surface rms of the corrected surface.**

With these definitions, we can define three measures of correctability as shown in Table 6.2.

Table 6.2 Correctability definitions.

CORRECTABILITY SYMBOL	VALUE	DEFINITION
C_{at}	$\frac{R_u - R_c}{R_u}$	**Total actuator correctability.**
C_{ab}	$\frac{R_b - R_c}{R_b}$	**Actuator correctability after perfect rigid-body motion removal.**
C_{af}	$\frac{R_f - R_c}{R_f}$	**Actuator correctability after perfect rigid-body motion and power removal.**

The importance of the differences between the correctability definitions listed in Table 6.2 is best illustrated by the following example.

6.2.2.1 Example: Adaptive control simulation of a mirror segment

The finite element model of a hexagonal mirror segment shown in Fig. 6.3 is to be used in an adaptive control simulation for two actuator configurations. The first actuator configuration shown in Fig. 6.3(a) consists of 19 compliant force actuators located at the open circles and three stiff-displacement actuators located at the solid circles. The second actuator configuration shown in Fig. 6.3(b) consists of 22 stiff-displacement actuators located at the solid circles. The

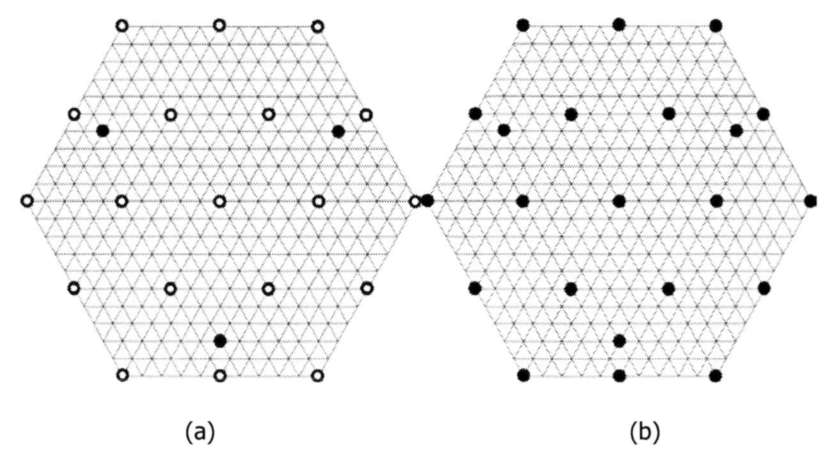

(a) (b)

Figure 6.3 Finite element models of a hexagonal mirror segment to be used in an adaptive control simulation: (a) first actuator configuration and (b) second actuator configuration. Solid circles indicate displacement actuator locations, while open circles indicate force actuator locations.

segment is subjected to a gravity load normal to its surface. The goal of the analysis is to predict the best-corrected surface and compute the various measures of correctability.

The uncorrected deformed surfaces due to the gravity load, influence functions of the center actuator, and corrected deformed surface are shown for each actuator configuration in Fig. 6.4, Fig. 6.5, and Fig. 6.6, respectively. The surface rms errors and correctabilities are shown in Table 6.3 for both actuator cases.

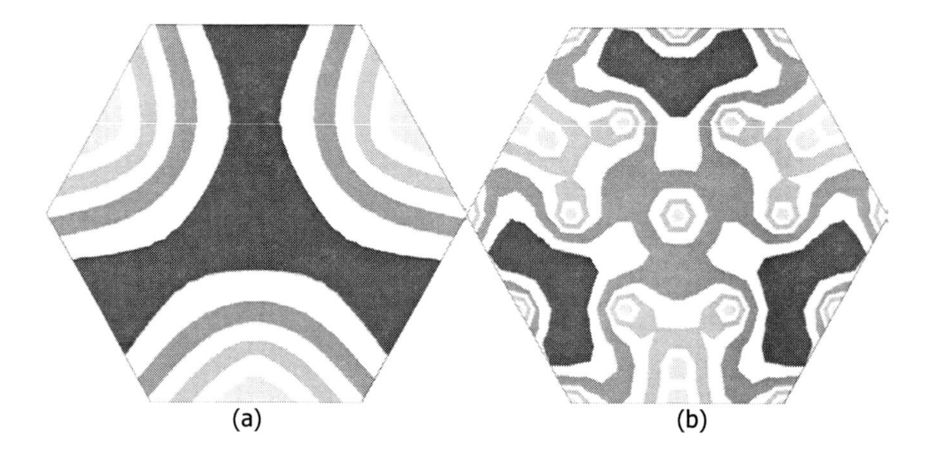

Figure 6.4 Uncorrected surface deformations due to gravity: (a) first actuator configuration and (b) second actuator configuration.

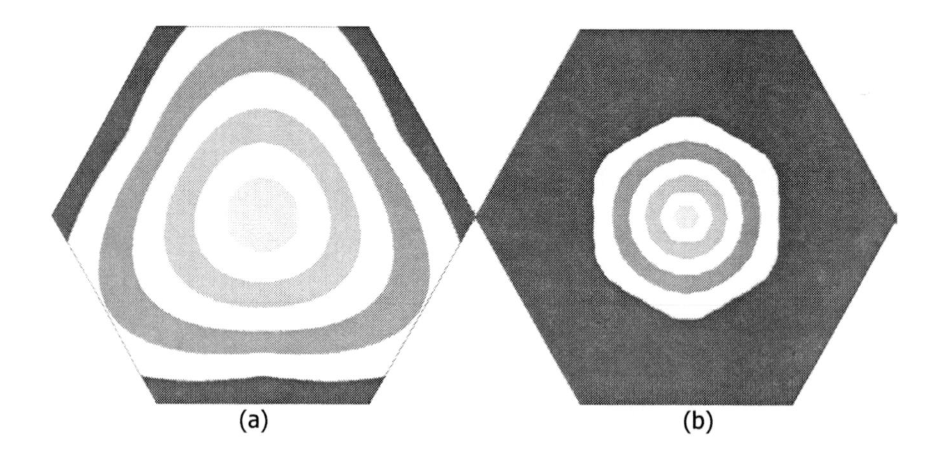

Figure 6.5 Influence functions of the center actuator: (a) first actuator configuration and (b) second actuator configuration.

Figure 6.6 Corrected surface deformation for both actuator configurations.

Table 6.3 Surface rms errors and correctabilities for adaptive hexagonal mirror.

RESULT TYPE	ACTUATOR CONFIGURATION 1	ACTUATOR CONFIGURATION 2
R_u	0.510 waves	0.021 waves
R_b	0.224 waves	0.008 waves
R_c	0.006 waves	0.006 waves
C_{at}	98.8%	72.4%
C_{ab}	97.4%	27.9%

Notice that even though the corrected surface rms error of each actuator configuration is the same, but the correctabilities are very different. This difference is attributed to the difference in the uncorrected surface error between the two designs. The correctability of the second actuator configuration design is lower than that of the second actuator configuration design only because its uncorrected performance is superior. What is most informative in this case from a design standpoint is that each design has the same corrected surface.

6.3 Coupled Adaptive Control Simulation and Structural Design Optimization

In the search for an adaptively controlled design that meets a set of requirements, it is valuable to use automated design optimization techniques. By simultaneously including both adaptive control and structural design variables, an automated optimization analysis can tune the structural design to the actuator layout and meet any other given requirements. Such an optimization analysis requires a calculation of the best correction to the surface error at each design cycle. However, many finite element software codes lack the ability to perform this calculation during the design optimization run. The method of modeling actuators, which is presented below, avoids this limitation.[2]

6.3.1 Method of modeling actuators in design optimization

Modeling of actuators in a finite element model is shown schematically in Fig. 6.7. The actuators are assumed to be stiff links that provide a variable-enforced relative motion between the optic and its mounting structure. Each actuator model consists of bar-element pairs connected to the same two nodes to which a change in temperature is applied. As shown in Fig. 6.7 the bars in each pair are given coefficients of thermal expansion, α, which are opposite in sign. The areas of each bar-element pair, A_1 and A_2, are driven in opposite directions by a single-design variable referred to as an actuator control variable. Therefore, any change in the actuator-control variable will cause the bar-element pair to grow or shrink as the bar with increasing area dominates over its respective counterpart whose area decreases. The sum of the areas is held at a relatively large constant so that the sensitivity of the optical surface rms deformation with respect to the actuator-control variable is significant enough to yield a well-posed optimization problem.

The expressions relating the bar areas to the actuator displacement are

$$A_1 = \frac{A_T \delta}{2\alpha} + \frac{A_T}{2}, \qquad [6.6(a)]$$

and

$$A_2 = -\frac{A_T \delta}{2\alpha} + \frac{A_T}{2}, \qquad [6.6(b)]$$

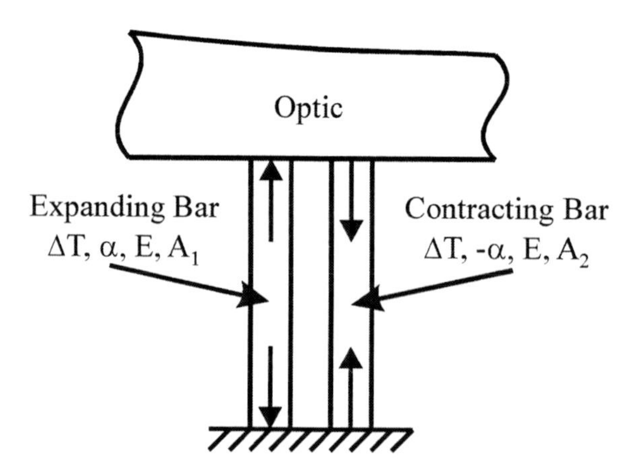

Figure 6.7 Schematic description of actuator model.

where A_1 and A_2 are the bar areas to be controlled by the actuator control variable, α is the coefficient of thermal expansion, L is the length, ΔT is an enforced change in temperature, E is the Young's modulus, and δ is the actuator displacement. These relations prescribe that in order to increase δ, A_1 must be increased by the same amount A_2 is decreased. In order to present the optimizer with a well- behaved design variable, the transformation

$$\delta = \frac{(x_a - b)}{C}, \tag{6.7}$$

is made from δ to the actuator control variable, x_a, where b is an offset, and C is a scaling factor. This transformation allows the actuator control variable, x_a, to range above and below a value, b, rather than zero. This avoids numerical difficulties that most optimizers experience with zero-valued design variables. It also allows control of the magnitude of the gradients of the objective function with respect to the actuator control variable through the scaling constant C.

The treatment of force actuators and displacement actuators is identical during any load case that computes the corrected surface as a design response. In other load cases or natural frequency computations in which it is desired that the force actuators be removed from the load path, a multipoint constraint, which is used to connect the force-actuator-bar pairs to the optic in the correctability load cases, can be turned off.

6.3.2 Guidelines for adaptive control design optimization

Although they are very powerful tools in engineering and other disciplines, numerical optimizers are unfortunately not entirely robust. The usefulness of the results is often highly dependent on the initial formulation of the design problem. These guidelines are given as general recommendations specific to optimization of adaptive optics in order to help the reader present the optimizer with a well-posed design problem capable of returning useful results. However, these observations are not complete in advising the reader on the avoidance of ill-posed formulations. Further reading of optimization techniques may provide additional insight in coaxing more complicated design problems to useful solutions.[3,4]

The combined optimization process involving both actuator-control and structural-design variables proceeds more efficiently if the initial values of the actuator-control variables are set to give the best correction to the optical surface for the initial structural design. When these initial actuator inputs are present, the optimizer can more easily assess how to improve the corrected surface by adjusting the structural-design variables. The actuator-control variables then only need to be given minor adjustments as the structural-design variables evolve. The initial values for the actuator-control variables can be found by performing an optimization on the initial structural design, including only the actuator-control variables.

When the optimizer is presented with initial-actuator inputs it is very important that the first search direction have significant components in the structural-design variables since consideration of only the actuator-control variables represents an optimally corrected surface. Therefore, it is important that the structural design variable to property relations and the shape basis vectors be scaled such that the objective sensitivities with respect to the structural-design variables are one or two orders of magnitude greater than the objective sensitivities with respect to the actuator-control variables. Failure to do this is evidenced by an optimization run that returns the initial structural-design variable values after one design iteration. Examination of the objective-gradient components will indicate how much the structural-design-variable-to-property relations must be scaled in order to generate a well-posed problem.

6.3.2.1 Example: Structural design optimization of an adaptively controlled optic

Figure 6.8 shows a half-symmetric finite element model of the initial design of a hexagonal adaptively controlled lightweight mirror fabricated of ULE. This mirror is to correct 2.0 HeNe waves of peak-to-valley surface error as defined by the $2\rho^2-1$-Zernike-polynomial term. This amount of power corresponds to a surface error of 0.52-HeNe-waves surface rms. The mirror is to be controlled by seven actuators whose locations are at the corners and center of the optic. A structural-design optimization that minimizes the corrected surface error is desired. Figure 6.9 shows a comparsion of the corrected surfaces for the optimized and unoptimized designs. While the unoptimized design corrects the surface error to 0.067-HeNe waves surface rms, the optimized design corrects the surface error to 0.023-HeNe waves surface rms. In addition, the optimized mirror offers a lower weight.

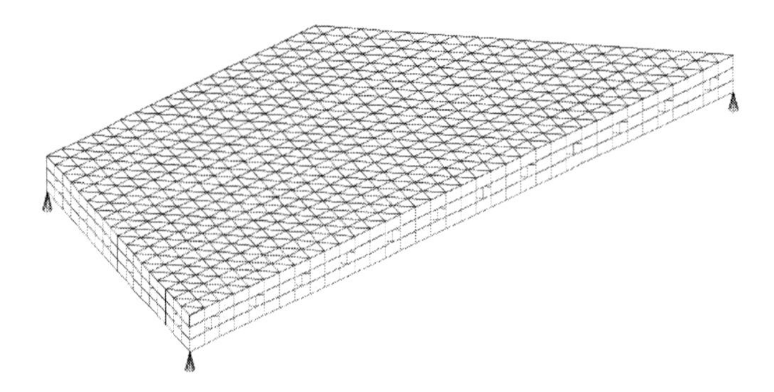

Figure 6.8 Finite element model of an adaptively controlled mirror to be optimized.

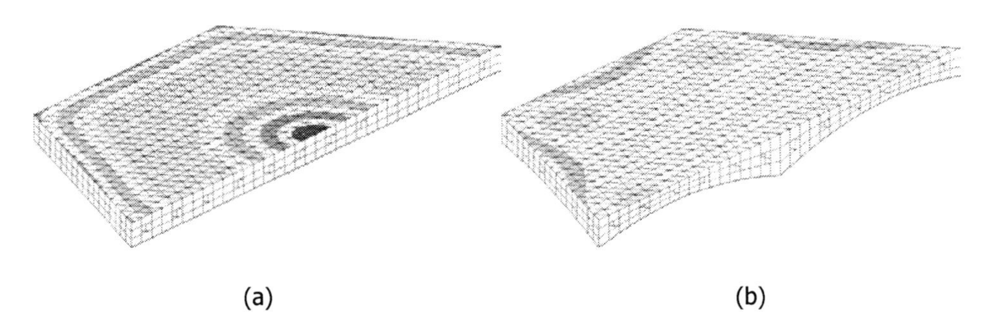

Figure 6.9 Corrected surfaces of constant depth and optimally sculpted adaptive mirrors: (a) constant depth mirror corrected to 0.067-HeNe-waves surface rms and (b) optimally sculpted mirror corrected to 0.023-HeNe waves surface rms.

References

1. Genberg, V., Michels, G., "Opto-mechanical analysis of segmented/adaptive optics," *Proceedings of SPIE*, **4444**, Bellingham, WA (2001).
2. Michels, G., Genberg, V., "Design optimization of actively controlled optics," *Proceeding of SPIE,* **4198**, Bellingham, WA (2000).
3. Vanderplaats, G., *Numerical Optimization Techniques for Engineering Design With Applications*, McGraw-Hill, New York (1984).
4. Onwubiko, C., *Introduction to Engineering Design Optimization*, Prentice Hall, Upper Saddle River, New Jersey (2000).

≺Chapter 7≻ Optimization of Optomechanical Systems

7.1 Overview

In a conventional design approach, the engineer cycles through trial designs until a satisfactory (feasible) design is found. In the unlikely case that there is still some budget and schedule available, the engineer may continue to run parametric studies to improve (optimize) the design. This is a trial-and-error effort requiring intuition and insight. If there are a significant number of design variables, the process is complex and time consuming.

Optimization theory offers some tools to improve the design process, including design sensitivity and nonlinear programming (NLP) techniques. When incorporated into a general purpose FEA program, optimization methods offer new opportunities for design improvement. In current software, the automated procedure will sequentially improve a starting design to obtain the "best" design. This "best" design is limited to the class of structures defined by the starting design and the choice of design variables. Because of the sequential nature of NLP, the "best" design may be a local minimum rather than a global minimum. Even with these caveats, optimization is a powerful tool in the hands of a knowledgeable user.

This chapter includes a brief overview of optimization theory; however, the main emphasis will be on the application of the tools to optomechanical systems. In a typical optical structure, many response quantities are calculated. These responses may have performance limits (constraints) applied, or may be optimized (objective).

ADVANTAGES OF OPTIMIZATION:

≺1≻ Provides logical, systematic, and complete design approach

≺2≻ Requires a total problem statement with all design requirements

≺3≻ Reduces design time; allows higher-level design trades

≺4≻ It generally works, since even a local optimum is an improvement.

DISADVANTAGES:

≺1≻ Requires computer tools, optimizer and compatible FE program

≺2≻ Requires knowledge of the tools and the theory

≺3≻ May get trapped in local optima

≺4≻ It may have difficulty with ill-posed problems.

TYPICAL RESPONSE QUANTITIES:

≺1≻ Structural: System weight, center-of-gravity, mass-moment-of-inertia
≺2≻ Structural: Stress, buckling, natural frequency, dynamic response
≺3≻ Optical: Image motion, jitter, MTF
≺4≻ Optical: Surface RMS after pointing and focus correction
≺5≻ Optical: Obscuration
≺6≻ Optical: OPD due to thermal or stress effects, birefringence
≺7≻ Optical: System wavefront error

Current technology allows for structural optimization using optical performance constraints (Sec. 7.3), or multidisciplinary thermal-structural-optical optimization (Sec. 7.4). This chapter does not address some other problems that could broadly fall under optomechanical design, such as optical beam path length optimization[1] in which optimization is used to solve a difficult geometry problem.

7.2 Optimization Theory

DEFINITIONS:

X = vector of design variables such as sizing, shape, material
R = vector of responses, typically nonlinear functions of X
F = objective function = a response to minimize or maximize
g = behavior constraint on a response as either an upper or lower bound

$$R \leq R^U \Rightarrow g = (R - R^U)/R^U \leq 0 \tag{7.1}$$

MATHEMATICAL DESIGN PROBLEM STATEMENT:

Minimize	$F(X)$		
subject to	$g \leq 0$	**behavior constraints**	
and	$X^L \leq X \leq X^U$	**side constraints.**	**(7.2)**

In this text, optimum design refers to the application of nonlinear programming techniques to find the best solution of the mathematical statement of the design problem.

If the design goal is to maximize the objective F, the problem can be stated in standard form by minimizing $-F$. If a response is limited by an equality constraint, it may be treated as two inequality constraints:

$$h = 0 \Rightarrow h \leq 0 \text{ and } h \geq 0. \tag{7.3}$$

Figure 7.1 shows a simple structural optimization of a three-bar truss with sizing variables (A_1, A_2, A_3) and shape variables (S, H). The objective is to minimize weight while satisfying performance constraints on displacement and stress and side constraints on size and shape.

The design space is an N-dimensional space with an axis for each of the N design variables, which is impossible to visualize if $N > 3$. A two-variable design space is depicted in Fig. 7.2. In most problems, the constraints are generally nonlinear functions of X and are often found numerically, which makes them expensive and difficult to plot, even in a 2D space. In the five-variable truss example, the stress and displacement are found via FEA, and all responses are nonlinear in S and H.

There is a variety of NLP techniques available[2] that move through the design space in a sequential manner. The most efficient techniques are gradient- based, requiring first derivatives (sensitivities) of the response quantities with respect to the design variables (dR/dX).

A common approach is to use finite differences to calculate sensitivities. Let X_0 represent a starting design point:

$$X_0 = (A_1, \ldots A_j, \ldots A_n), \tag{7.4}$$

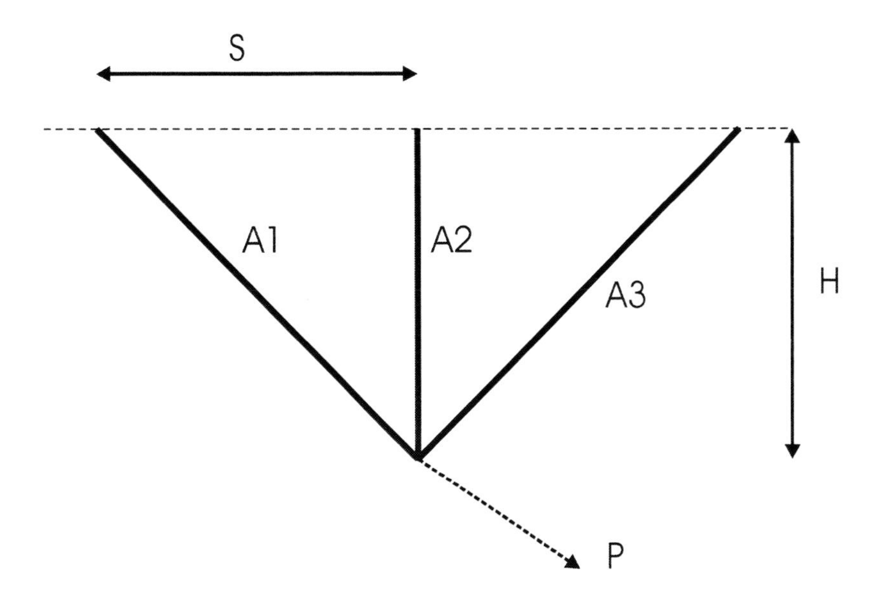

Figure 7.1 Three-bar truss.

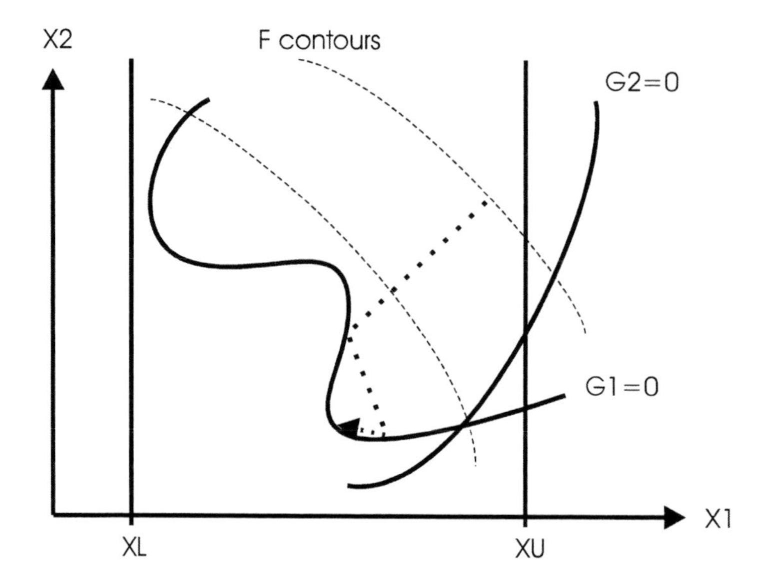

Figure 7.2 Two-variable design space.

which is evaluated via FEA:

$$K_0U_0 = P_0 \Rightarrow U_0. \tag{7.5}$$

The derivative of displacement with respect to design variable A_j is found by perturbing the design:

$$X_j = (A_1,...A_j + \Delta A_j,...A_n), \tag{7.6}$$

then reevaluating with FEA:

$$K_jU_j = P_j \Rightarrow U_j, \tag{7.7}$$

and computing a finite difference derivative:

$$U_j' = dU/dX_j = (U_j - U_0)/\Delta A_j. \tag{7.8}$$

This is a very general technique, but quite expensive computationally.

A more efficient technique uses implicit derivatives of the initial equation [Eq. (7.5)]:

$$K_0U' + K'U_0 = P'. \tag{7.9}$$

The derivative U' can be solved from

$$K_0 U' = P' - K' U_0 = P^*, \tag{7.10}$$

which is the equivalent cost of an additional load case P^* in the original solution. Note that K' and P' are relatively cheap to calculate in many cases. For the example truss problem:

$$k = AE / L \Rightarrow k' = dk / dA = E / L. \tag{7.11}$$

For external forces, p' is 0. For a gravity body force,

$$p = AL\rho g / 2 \Rightarrow p' = dp / dA = L\rho g / 2. \tag{7.12}$$

Most other responses can then be found from U' by the chain rule. For example, the stress sensitivity in the truss is found from

$$d\sigma / dX = (d\sigma / dU) U' \Rightarrow d\sigma / dU = E / L. \tag{7.13}$$

Typical design optimizations require more than 100 design cycles to optimize. For large models, the computer time for 100 analyses is excessive. A huge efficiency can be gained by using the design sensitivities and approximation theory[2] to create a response surface. The steps in this approach are

≺1≻ given a design X_q at design cycle q,
≺2≻ run a full FE analysis along with design sensitivity,
≺3≻ create approximate problem (response surface) via Taylor series

$$g^* = g(X_q) + g'(X_q) / (X - X_q), \tag{7.14}$$

≺4≻ optimize the approximate problem very quickly to get X_{q+1},
≺5≻ check convergence; loop back to step 1.

In this approach (shown in Fig. 7.3), step 4 requires hundreds of cheap optimizations, while the expensive FEA in step 2 is typically 10–20 analyses. Commercial FEA programs using this approach include MSC/Nastran and VRD/Genesis.

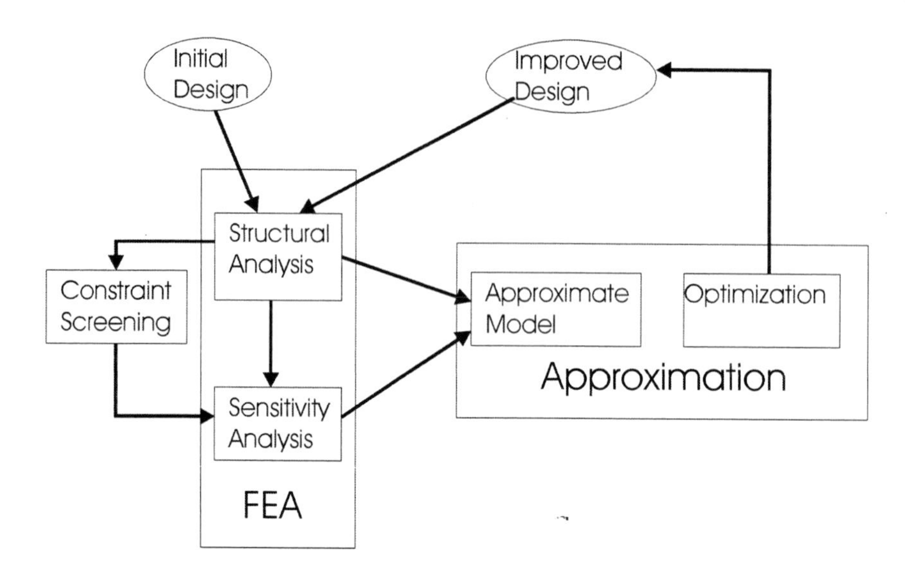

Figure 7.3 Optimization flow using approximation theory.

7.3 Structural Optimization, Including Optical Measures

In this section, it is assumed that the optical design is fixed, leaving only the structural variables to design. Commercially available software allows some of the optical response quantities to be incorporated into the structural FEA optimization model.[3–6] The specific capabilities listed in this section are found in MSC/Nastran and Sigmadyne/SigFit[7] software packages.

The most easily incorporated optical criterion is image motion and defocus. In Chapter 4, sample equations of image motion for a simple telescope are presented as multipoint constraint (MPC) equations. For small motion, these are linear equations that could be added into any FEA code allowing linear equation input. If an optic, such as the primary mirror, is deforming, then some average surface motion must be calculated for each of the optical terms. In Nastran, an RBE3 can approximate the average motion without affecting the stiffness. A better representation of surface tilts, bias, and even radius-of-curvature are calculated within SigFit and written to Nastran MPC format. (See Zernike fit below.)

Wavefront error budgets typically specify a surface rms or peak-to-valley (P-V) requirement for surface distortion under a variety of test and in-use load conditions. These budgets usually require that the pointing and focus terms are subject to one error budget and that the residual surface distortion is subject to a separate wavefront error budget. This is accomplished in SigFit by writing the Zernike polynomials as MPC equations,[5,8] subtracting the tilt, bias, and focus terms, and then calculating the residual rms or P-V using Nastran's nonlinear equation feature. The procedure is outlined as follows:

U_k = displacement of node k from finite element solution,
$C_j = j^{\text{th}}$ Zernike coefficient,
F_{jk} = node k displacement due to unit value of jth Zernike;
$Z_k = \Sigma_j C_j F_{jk}$ = Zernike representation of node k displacement. (7.15)

When fitting coefficients C to a deformed shape U, the error E, where

$$E = \Sigma_k W_k (U_k - Z_k)^2 \tag{7.16}$$

is minimized with respect to the coefficients

$$dE / dC_j = 0, \tag{7.17}$$

resulting in a linear system:

$$[H]\{C\} = \{R\}. \tag{7.18}$$

Solving for C,

$$\{C\} = \left[H^{-1}\right]\{R\} = [A]\{U\}, \tag{7.19}$$

which can be represented as MPC equations with the C as SPOINTS. Another set of MPC equations subtract the user-selected Zernike terms and place the residual error into a dummy surface mesh:

$$E_k = U_k - \Sigma_j C_j F_{jk}. \tag{7.20}$$

The residual error rms or P-V is calculated using DRESP2 and DEQATN entries:

$$\text{rms} = \text{SQRT}[\Sigma_k W_k E_k{}^2], \tag{7.21}$$

$$\text{P} - \text{V} = \max(E_k) - \min(E_k). \tag{7.22}$$

Any of the above responses may be treated as constraints or as the objective in the Nastran optimization solution. The telescope example in the next chapter shows the application of these techniques.

For optical performance quantities that are not easily represented as bulk data equations, a new feature is available in MSC/Nastran called DRESP3. This allows an external program to be executed to calculate a response quantity in the optimization loop for use in constraint or objective calculations. For example, SigFit could be called to calculate the surface rms due to random loading or MTF due to vibration. Another scenario might have SigFit process the Nastran analysis, then call ORA/CODEV to evaluate the optical performance, returning any metric that ORA/CODEV calculates.

When optimizing lightweight mirrors, the 2D and 3D equivalent models in Chapter 3 are especially useful in preliminary design. Important design quantities such as cell size (B) are easily incorporated as design variables in an equivalent stiffness design, yet impossible to modify in a full 3D model. For a mirror with symmetric front and back plate thickness and a hexagonal core, a typical design flow is to find B, T_p, T_c, H_c from a 2D or 3D equivalent stiffness model.

DESIGN VARIABLES FOR LIGHTWEIGHT MIRROR = B, T_p, T_c, H_c:

B = cell-size (inscribed circle diameter)
T_p = front and back faceplate thickness
T_c = core wall thickness
H_c = core height.

2D EQUIVALENT STIFFNESS (ALL PSHELL-SIZE VARIABLES):

$H = H_c + 2T_p$ =core height
$\alpha = T_c/B$ = core solidity ratio
$T_m = 2T_p + \alpha H_c$, = membrane thickness (size variable)
$I_b = [H^3 - (1-\alpha)H_c^3] / 12$ = bending inertia
$R_b = 12I_b / T_m^3$ = bending ratio (size variable)
$S = [H^2 - (1-\alpha)H_c^2] / \alpha$ = transverse shear
$R_s = 8I_b/ST_m$ = bending ratio (size variable)
NSM = $\alpha H_c \rho$ = additional nonstructural mass.

3D EQUIVALENT STIFFNESS MODEL (SIZE, SHAPE, AND MATERIAL VARIABLES):

$\alpha = T_c/B$ = core solidity ratio
$E^* = \alpha E$ = effective modulus of core (material variable)
$\rho^* = 2\alpha\rho$ = effective density of core (material variable)
DVGRID = moves grid position for H (shape variable)
T_p = PSHELL faceplate thickness (size variable).

The 1g or polishing-quilting effects can be included in (a) or (b) as design constraints by using the quilting equations presented in Chapter 3. For complex cell geometry involving cathedral ribs, a separate breakout model of a single cell can be included in the overall optimization model. The results of 2D or 3D equivalent stiffness optimization are used to create a full 3D model, which can then be optimized again to refine values for T_p, T_c, and H_c for additional 3D effects. An example of a mirror optimization appears in the telescope example in Chapter 8.

Optimization can be combined with adaptive optics theory in Chapter 6 to improve the design of adaptive lightweight mirrors.[9] Although this approach improved the structural design, the number and placement of actuators were not treated as design variables. Recent advancements suggest that actuator number and placement will soon be allowed as design variables.

7.4 Integrated Thermal-Structural-Optical Optimization

Without the ability to work concurrently, the disciplines of thermal control, structural design, and optical systems design levy worst-case performance requirements on each other such that each specialty can contribute to a design independently. In a typical design approach, the optical engineer creates a performance error budget that dictates distortion limits to the structural engineer, who then dictates limits on temperatures and gradients to the thermal engineer. Requirements are derived, then flowed down (Fig 7.4). The thermal and structural engineers attempt to achieve these limits under all operational conditions. Such an approach satisfies the optical performance requirements, but at a cost of overdesign due to stacked-up margins. For example, temperature gradients in a mirror support structure are inconsequential as long as the required optical performance is achieved yet derived limits on such gradients often become a design driver for thermal control specialists. To achieve higher performance, an integrated design approach based on multidisciplinary design optimization (MDO) is required.

OptiOpt[10] is an integrated package for designing and optimizing high-performance optical systems using MDO methods. This NASA-funded SBIR project achieved the integration of specialized design tasks such as thermal and structural integrity verification into a design optimization environment that met mission optical performance specifications, avoiding over design by eliminating derived or flowed-down requirements. The approach is unique in that it respects the models and approaches of each engineering specialty (thermal, structural, and optical), and, most importantly, it preserves the use of the design tools most commonly used by these specialties.

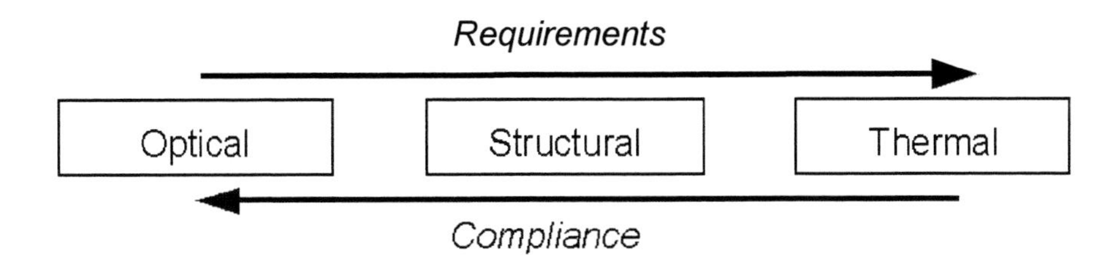

Figure 7.4 Requirements flow in optomechanical design.

In OptiOpt, the following popular and commercially available programs were integrated:

<1> Engineous/Isight for optimization,
<2> ORA/CODEV for optical analysis,
<3> MSC/Nastran for structural analysis,
<4> CRT/ThermalDesktop for thermal analysis,
<5> Sigmadyne/Nascode* for structural-optical I/F.

Figure 7.5 shows the overall data flow during the design optimization. Reference [10] uses the telescope model in Chapter 8 as a candidate for design optimization. In the first telescope design optimization using low-CTE materials (Invar and ULE), an optimized design meeting all performance requirements weighed 155 pounds. In the second optimization, the metering structure and primary mirror were switched to lower-cost but higher-CTE materials (aluminum and fused silica), providing another feasible optimum of even lighter weight at 66 pounds. The second design required more thermal control to meet performance requirements, driving the total package weight to 189 pounds.

In the two optimizations discussed above, the optical design variables (diameter, radius-curvature, spacing) were fixed throughout. A more interesting optimization has yet to be run, in which the optical design quantities are included as design variables. The fully optimized design offers greater potential for weight reduction. Other studies[11] have confirmed that MDO offers significant performance improvement.

As with any MDO, computer resources are stretched to their limits. To make MDO more efficient, the advances achieved in structural optimization must be applied to the other disciplines. A key to efficiency is the development of interdisciplinary design sensitivity derivatives for use in the response surface as noted in the previous section. Of course, this requires an optimization routine such as DOT[12] that directly accepts sensitivity information rather than just function evaluations.

---

* Sigmadyne/Nascode has been replaced by an enhanced Sigmadyne/SigFit for thermal-structural-optical I/F.

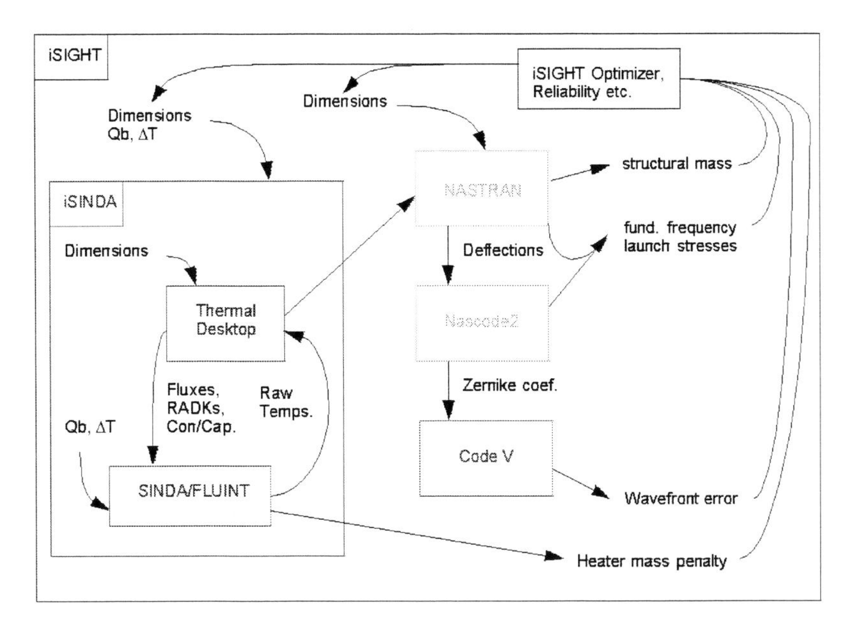

Figure 7.5 OptiOpt data flow.

References

1. Genberg, V., "Beam pathlength optimization," *Proceedings of SPIE*, **1303**, Bellingham, WA (1990).
2. Vanderplaats, G. *Numerical Optimization Techniques for Engineering Design, 3rd Ed.*, VR&D, (1999).
3. Genberg, V., Cormany, N., "Optimum design of lightweight mirrors," *Proceedings of SPIE*, **1998** (1993).
4. Thomas, H., Genberg, V., "Integrated structural/optical optimization of mirrors," *Proceedings of AIAA*, 94-4356CP (1994).
5. Genberg, V., "Optimum design of lightweight telescope," *Proceedings of MSC World Users Conference* (1995).
6. Genberg, V., "Optical performance criteria in optimum structural design," *Proceedings of SPIE*, **3786**, (1999).
7. *SigFit* is a product of Sigmadyne, Inc., Rochester, NY.
8. Genberg, V., "Optical surface evaluation,"*Proceedings of SPIE*, **450** *Structural Mechanics in Optical Systems*, (1983).
9. Michels, G., Genberg, V.,"Design optimization of actively controlled optics," *Proceedings of SPIE,* **4198**, (2000).
10. Cullimore, Panczak, Bauman, Genberg, Kahan, "Automated multi-disciplinary optimization of a space-based telescope," *Proceedings of ICES* (2002).
11. Williams, Genberg, Gracewski, Stone, "Simultaneous design optimization of optomechanical systems," *Proceedings of SPIE*, **3786**, (1999).
12. Vanderplaats, G., *DOT Users Manual*, VR&D (1999).

≺Chapter 8≻ Integrated Optomechanical Analysis of a Telescope

8.1 Overview

A simple two-mirror telescope will be used to demonstrate integrated analysis techniques common in optomechanical systems. Although this design was created for analysis demonstration only, the performance requirements placed on this system are representative of real applications. The level of detail in these models is quite coarse yet consistent with a conceptual design study model; it also keeps the model files small and readable. The models are provided on a CD that accompanies this text; they are in MSC/Nastran and ORA/CODEV format, since both programs represent market leaders in their segment. Most FE preprocessors can read Nastran data files and convert to other FE codes. The Readme.txt file on the CD explains the filenames used for each analysis described below.

IN THIS CHAPTER, THE FOLLOWING NOTATION WILL BE USED EXTENSIVELY:

PM = primary mirror (just the optic)
PMA = primary mirror assembly (optic plus mount)
SM = secondary mirror (just the optic)
SMA = secondary mirror assembly (optic plus mount)
FP = focal plane
IM = image motion.

8.2 Optical Model Description

The primary mirror (PM) and secondary mirror (SM) are both centered on the *z*-axis, as is the detector, with the global origin at the PM vertex. The local coordinate systems for the PM, SM, and FP are right-handed and identical except for their *Z* positions (see Fig. 8.1). The optical prescription is given in Table 8.1.

Table 8.1 Optical prescription.

	RADIUS	CONIC CONSTANT	AXIAL (Z) POSITION
PM	$R_1 = -101.5332''$	−1.0023	0
SM	$R_2 = -12.4893''$	−1.4969	−45.1204″
FP	flat	NA	+13.7922″

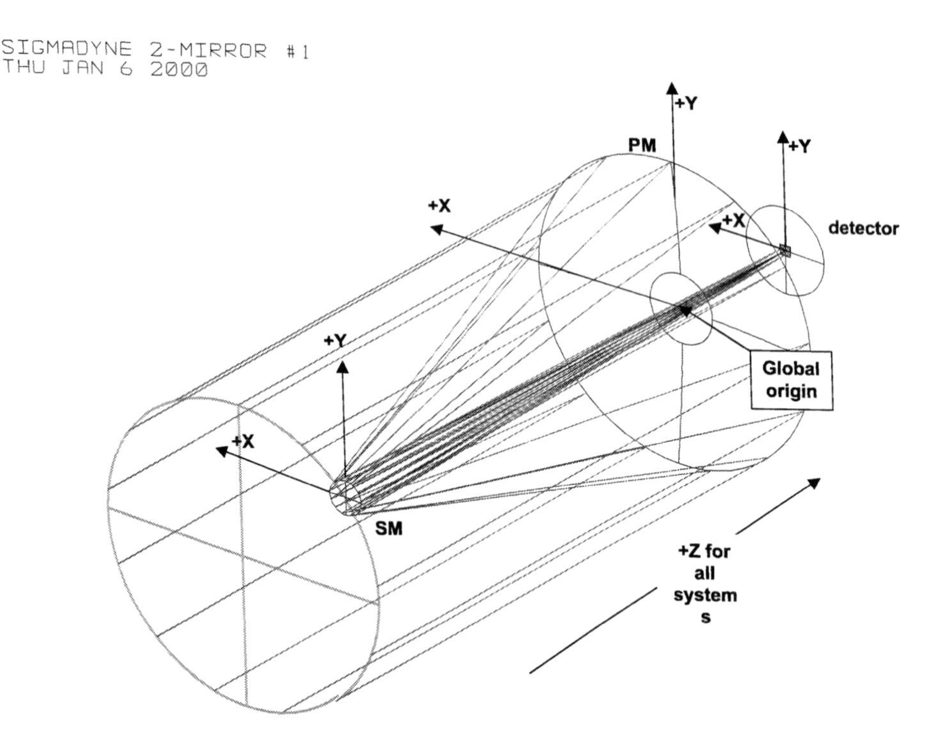

Figure 8.1 Telescope optical layout.

To be consistent with the structural modeling techniques, global coordinates were used in the CodeV model to define the optical surfaces, rather than sequential surface-to-surface definition.

8.3 Structural Model Description

A cutaway plot of the finite element structural model is shown in Fig. 8.2. In this model, the shell and spider are invar. The PM is ULE and represented as a 3D equivalent stiffness model on three bipod flexures of titanium. The SM is solid ULE and held by three titanium blade flexures. The focal plane is supported by an invar spider. Since this model represents a very early concept model, the adhesive bonds for the PM and SM are not included. Therefore, bond effects are not predicted by this model, but would be included in later studies.

Coordinate systems in the structural model match the coordinate systems in the optical file. The MSC/Nastran model of the telescope is subdivided into components for easy manipulation. The data files are heavily commented so that a reader can follow input and modify as desired.

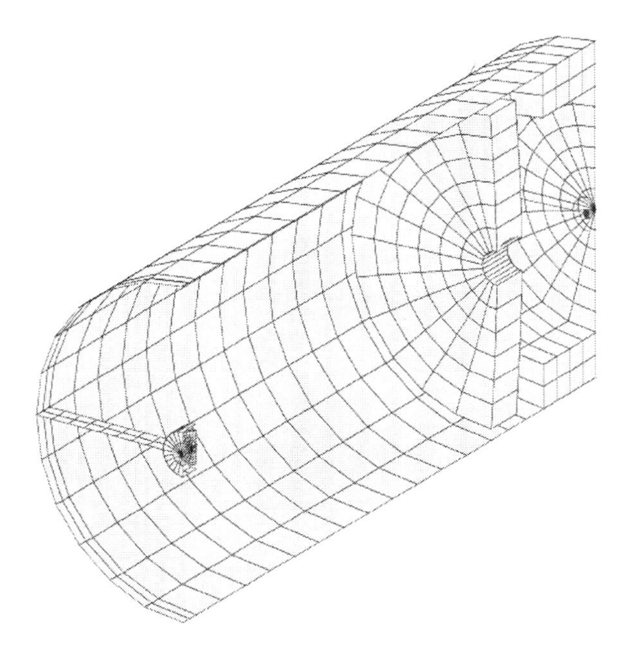

Figure 8.2 Structural model cut-away view.

8.4 One-Gravity Static Performance

Since all systems must be tested in a 1-g environment, the predicted 1-g static performance in typical test orientations must be analyzed. The Nastran FEA results for various orientations were analyzed in SigFit to create Zernike coefficient tables and CODEV files for rigid body motion (.seq files) and Zernike coefficients (.int files). The results for PM due to 1-g +Z are given in Table 8.2.

The raw finite element displacements are dominated by rigid body effects. If rigid body motion is subtracted, the resulting surface is shown in Fig. 8.3 on the FE model. For further optical processing, the FE data may be interpolated in SigFit[1] to a rectangular array in interferogram format as shown in Fig. 8.4. This data may be directly compared to experimental interferograms for correlation, and they can be used as "backouts" to subtract 1-g effects for on-orbit predictions[2].

8.5 On-Orbit Image Motion Random Response

The image-motion-sensitivity matrix in image space (Table 8.3) was determined by perturbing each of the optics a small amount in each coordinate direction. These coefficients were then written in Nastran MPC format for inclusion into a dynamic analysis.

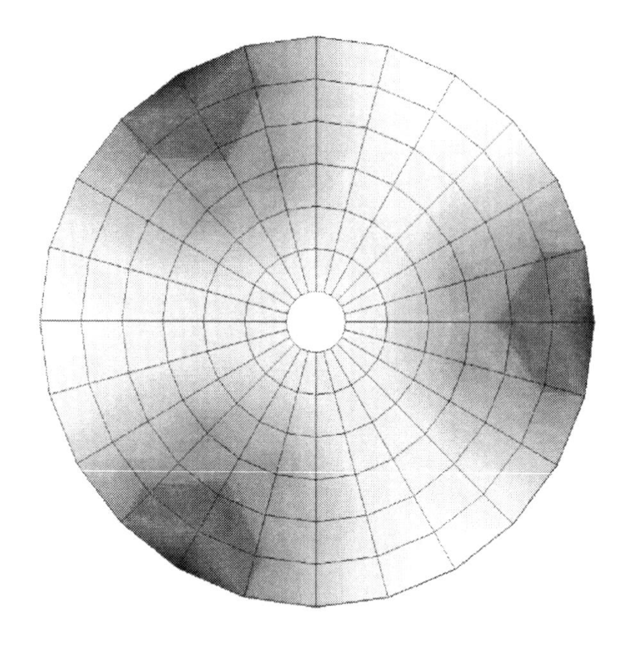

Figure 8.3 PM Surface with Rigid-Body Removed for 1g +Z.

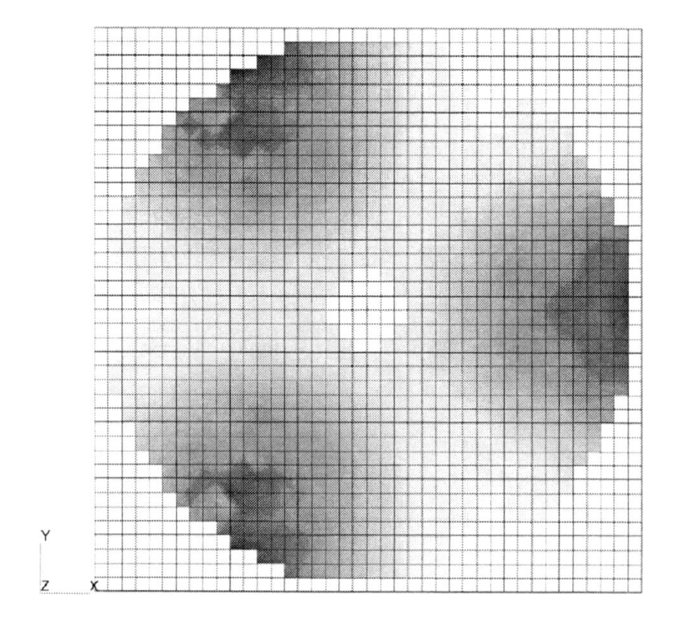

Figure 8.4 Interferogram Array of PM for 1g +Z.

Table 8.2 Zernike fit of PM for 1-g +Z.

Sigmadyne, Inc. SigFit Version = v2002-r2 28-Jun-02 11:16:22
EXAMPLE TELESCOPE MODEL
INVAR & ULE
1G +Z
INPUT SURFACE: Best-Fit Zernike Polynomial Coefficients

Loadcase = 3	**Subcase Id = 3**	**LoadFactor = 1.0000E + 00**
Surface = 1	**FE units = in**	**Wavelength = 2.3622E – 05**
Optic-Id = 2	**OpticLabe l = "PM"**	

ORDER K	N	M	ABERRATION	MAGNITUDE (WAVES)	PHASE (DEG)	RESIDUAL RMS	RESIDUAL P-V
Input (wrt zero)						.1051	.4477
1	0	0	Bias	–.00060	.0	.1051	.4477
2	1	1	Tilt	.00000	.0	.1051	.4477
3	2	0	Power (Defocus)	.03976	.0	.1029	.4477
4	2	2	Pri Astigmatism	.00000	.0	.1029	.4477
5	3	1	Pri Coma	.00000	.0	.1029	.4477
6	3	3	Pri Trefoil	.26430	.0	.0392	.1757
7	4	0	Pri Spherical	–.03348	.0	.0363	.1409
8	4	2	Sec Astigmatism	.00000	.0	.0363	.1409
9	4	4	Pri Tetrafoil	.00000	.0	.0363	.1409
10	5	1	Sec Coma	.00000	.0	.0363	.1409
11	5	3	Sec Trefoil	.09301	–60.0	.0245	.1527
12	5	5	Pri Pentafoil	.00000	.0	.0245	.1527
13	6	0	Sec Spherical	.01920	.0	.0232	.1249
14	6	2	Ter Astigmatism	.00000	90.0	.0232	.1249
15	6	4	Sec Tetrafoil	.00000	.0	.0232	.1249
16	6	6	Pri Hexafoil	.02965	.0	.0213	.1066
17	7	1	Ter Coma	.00000	.0	.0213	.1066
18	7	3	Ter Trefoil	.03486	.0	.0174	.0719
19	7	5	Sec Pentafoil	.00000	.0	.0174	.0719
20	8	0	Ter Spherical	–.00646	.0	.0171	.0722
21	8	2	Qua Astigmatism	.00000	90.0	.0171	.0722
22	8	4	Ter Tetrafoil	.00000	45.0	.0171	.0722
23	8	6	Sec Hexafoil	.00132	.0	.0171	.0716

Rmax: Normalizing Radius (FE units) = 1.4000E+01.
Polynomials normalized to have unity magnitude at Rmax.
Note: Polynomials defined so +bump is in -Z.
Fit: Surface Normal Displacement (dn) vs Radial position (r).
Displ: Decenters ARE subtracted prior to Polynomial fit.
Displ: BFP subtracted prior to Polynomial fit.
Rigid Body Tx,Ty,Tz (FE units) 6.0978E-13 5.0466E-17 2.5604E-04.
Rigid Body Rx,Ry,Rz (Radians) -3.4014E-19 -5.3029E-13 -6.3114E-21.
Orig-RoC,dRoC,dCoC (FE units) -1.0153E+02 5.9365E-04 5.9565E-04.
RMS and Peak-to-Valley (P-V) are calculated on the residual surface after each polynomial term subtracted in the order listed above.
Above fit for deformed SURFACE, for reflected WAVEFRONT multiply by 2.

Table 8.3 Image motion sensitivity matrix.

OPTIC MOTION	IMAGE X	IMAGE Y
PM Rotation about X	0.000	1059.400
PM Rotation about Y	–1059.400	00.000
PM Displacement X	10.430	40.000
PM Displacement Y	0.000	10.434
SM Rotation about X	0.000	117.830
SM Rotation about Y	117.830	0.000
SM Displacement X	–9.434	0.000
SM Displacement Y	0.000	–9.434
FP Displacement X	–1.000	0.000
FP Displacement Y	0.000	–1.000

Table 8.4 Base shake PSD input to mounts.

FREQ (HZ)	PSD (G^2/HZ)
2.00000E + 01	2.00000E-08
6.00000E + 01	2.00000E-06
1.40000E + 02	2.00000E-06
1.80000E + 02	2.00000E-08

For dynamic excitation, a "base shake" input (Table 8.4) was applied through the main telescope mounts to represent on-board disturbances. A random response analysis was run in Nastran to calculate the 1σ response of the IM as 0.00315 inches. The Nastran model could not predict the random response of the surface distortion, represented as surface rms.

To obtain more information about the dynamic response, a SigFit analysis was conducted using Nastran calculated natural frequencies and mode shapes. A random response analysis was conducted in SigFit to determine image motion response and the surface rms response. The analysis output presents the image motion jitter response and identifies the key modal contributors[3] to that response in Table 8.5. Similarly, the PM average rigid body motions and surface rms are presented in Table 8.6 with their key modal contributors in Table 8.7. These key modes could now be investigated by plotting strain energy density to see if design improvements could reduce the response. In addition, the key mode shapes could be fit with Zernike coefficients to pass to CodeV for detailed prediction of optical performance. The image motion results could be converted into jitter MTF using the tools in Chapter 4.

Table 8.5 Modal contributors to image motion.

MODE	FREQ	RB-TX	RB-TY
4	70.07	0.000	0.000
5	73.77	1.074	49.693
6	73.77	26.756	49.693
7	79.29	0.000	.307
8	79.29	40.418	.307
9	82.11	0.000	0.000
10	82.92	0.000	0.000
11	85.29	0.000	0.000

Table 8.6 Random response PSD of PM.

RANDOM ANALYSIS RESULTS:
1-SIGMA => contains peaks 68.3% time
3-SIGMA => contains peaks 99.7% time
ZERO-XS => number zero crossings/unit time
RIGID BODY TRANSLATIONS (FE UNITS): RB-Tx,...
RIGID BODY ROTATIONS (RADIANS): RB-Rx,...
S-RMS = Surface rms AFTER BFP removed (Waves)

SURF#	ITEM	1-SIGMA	3-SIGMA	ZERO-XS
1	RB-Tx	4.8898E-05	1.4669E-04	1.1174E+02
1	RB-Ty	6.7354E-13	2.0206E-12	9.6182E+01
1	RB-Tz	2.6656E-13	7.9968E-13	1.1777E+02
1	RB-Rx	4.9417E-14	1.4825E-13	1.3178E+02
1	RB-Ry	2.4892E-06	7.4677E-06	9.9650E+01
1	RB-Rz	7.7558E-15	2.3267E-14	1.2533E+02
1	d-RoC	7.5254E-10	2.2576E-09	1.4158E+02
1	S-RMS	1.7303E-02	5.1908E-02	1.3948E+02

Table 8.7 Modal contributors to PM PSD.

EACH MODES PERCENTAGE CONTRIBUTION TO PSD FOR PRIMARY MIRROR

MODE	FREQ	RB-TX	RB-TY	RB-TZ	RB-RX	RB-RY	D-ROC	S-RMS
4	70.07	0.000	0.000	0.000	0.000	0.000	0.000	0.000
5	73.77	3.654	49.988	1.023	44.967	0.141	4.608	15.827
6	73.77	90.993	49.988	0.054	44.967	3.520	87.158	78.981
7	79.29	0.000	0.012	0.000	5.029	0.000	0.063	0.004
8	79.29	5.337	0.012	0.096	5.029	96.281	8.072	5.001
9	82.11	0.000	0.000	0.000	0.000	0.000	0.000	0.000
10	82.92	0.000	0.000	0.000	0.000	0.000	0.000	0.000
11	85.29	0.000	0.000	0.000	0.000	0.000	0.000	0.000
12	85.29	0.000	0.000	0.000	0.000	0.000	0.000	0.000
13	90.31	0.000	0.000	0.000	0.000	0.000	0.000	0.000
14	90.31	0.001	0.000	0.001	0.000	0.000	0.000	0.002

8.6 Optimizing PMA with Optical Measures

To reduce the weight of the primary mirror, an optimization using optical measures[4] was performed using the techniques in Sect. 7.3. During the concept design phase, the mirror and its bipod mount were optimized as an assembly.

DESIGN VARIABLES FOR LIGHTWEIGHT MIRROR = B, T_P, T_C, H_C

- **B = cell-size (inscribed circle diameter)**
- **T_p = front and back faceplate thickness**
- **T_c = core wall thickness**
- **H_c = core height**

DESIGN VARIABLES FOR THE BIPOD MOUNT = D, L, A

- **D = diameter of flexure**

The Nastran control file contains load cases for launch loads with stress constraints, 1g test requirements, and a subcase with natural frequency constraints. Polishing quilting and 1-g quilting requirements are included as equations. The 1-g test requirement is that the surface rms be less than a specified value after pointing and focus correction (i.e., after bias, tilt, and power are removed). The equations for calculating the surface rms after correction were written in Nastran format (MPC, DRESP2, DEQATN) by SigFit. Table 8.8 lists the initial and final design values for variables and responses.

Table 8.8 PMA optimization results.

DESIGN VARIABLES	START	END	
FacePlate Thickness	0.18	0.10	
Core Thickness	0.08	0.06	
Core Height	3.80	4.07	
Cell Size	4.00	3.00	
Strut Diameter	0.05	0.74	
RESPONSES	**START**	**END**	
Weight	24.50	16.30	Minimize
Surface RMS	0.100	00.92	Limit <0.92
Quilting	00.72	00.75	Limit <0.75
Natural Frequency	47.00	85.00	Limit >60
Strut Stress	43000.00	13000.00	Limit <60,000

Once a suitable cell size (B) is chosen, a full 3D model can be created for further optimization with more detailed design criterion. With actual cell geometry modeled, quilting is predicted directly by the model, eliminating the need for quilting equations.

8.7 Adaptive PM

During the concept trades, an adaptive PM was considered as an alternative for reducing mirror weight. A much thinner mirror with six force actuators was studied. The 2D equivalent stiffness model along with the six unit actuators in a "bed-of-nails" configuration is shown in Fig. 8.5. The actuators may be used to correct the surface for a variety of in-use and test conditions.[3] In this example, a 1g +*Z* test configuration is presented.

Although this actuator scheme used six actuators, there were no rigid body displacement actuators present to control bias, tip, and tilt. Within SigFit, the force actuators were augmented (by user selection) with three rigid body actuators to account for pointing control elsewhere in the system.

For the thin PM in a 1-g +Z environment, the uncorrected and corrected surfaces are shown in Figs. 8.6 and 8.7, respectively. Optimization of adaptive mirrors is discussed in Ref. 5.

8.8 System-level Multidisciplinary Optimization

The telescope discussed in this chapter is the same telescope used for an MDO study as described in Sec. 7.4.

Although this actuator scheme used six actuators, there were no rigid body displacement actuators present to control bias, tip, and tilt. Within SigFit, the force actuators were augmented (by user selection) with three rigid body actuators to account for pointing control elsewhere in the system.

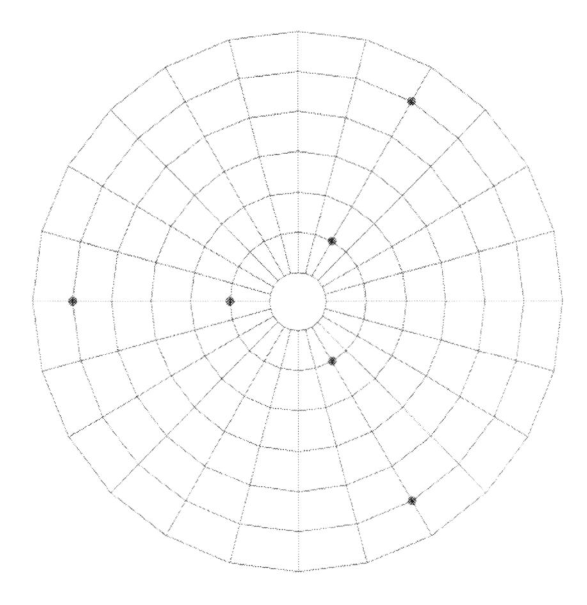

Figure 8.5 Adaptive primary mirror.

For the thin PM in a 1-g +Z environment, the uncorrected and corrected surfaces are shown in Figs. 8.6 and 8.7, respectively. Optimization of adaptive mirrors is discussed in Ref. 5.

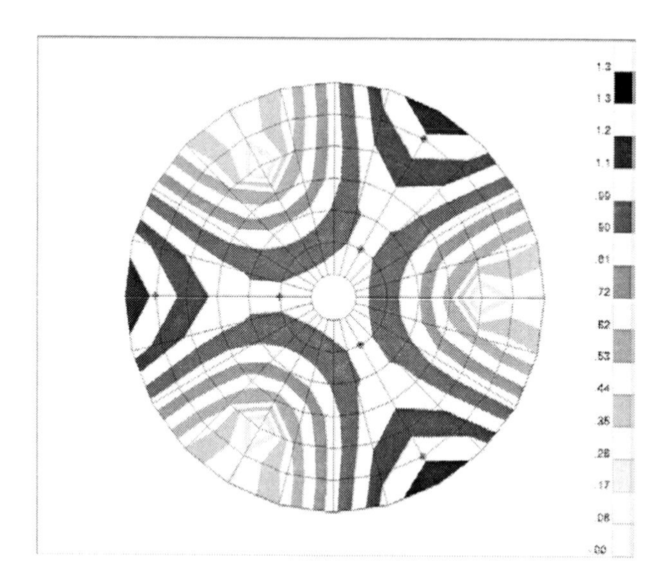

Figure 8.6 Uncorrected 1g Distortion (rms=0.30 λ).

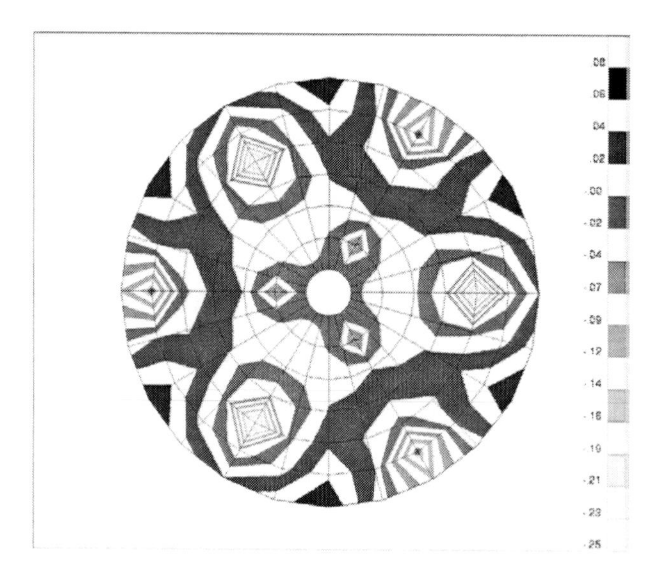

Figure 8.7 Corrected 1g Distortion (rms = 0.06 λ).

References:

1. SigFit Reference Manual, Sigmadyne Inc., Rochester, NY (2002).
2. Genberg, V., Michels, G., Doyle, K., "Making mechanical FEA results useful in optical design," *Proceedings of SPIE*, **4761**, Bellingham, WA, (2002).
3. Genberg, V., Michels, G., "Opto-mechanical analysis of segmented/adaptive optics," *Proceedings of SPIE*, **4444**, Bellingham, WA (2001).
4. Genberg, V., "Optical performance criteria in optimum structural design," *Proceedings of SPIE*, **3786** (1999).
5. Michels, G., Genberg, V.,"Design optimization of actively controlled optics," *Proceedings of SPIE*, **4198** (2000).

≺Chapter 9≻ Integrated Optomechanical Analysis of a Lens Assembly

9.1 Overview

Thermal, structural, and optical modeling tools are used to predict the optical performance of a Double Gauss lens assembly subject to laser loading as shown in Fig. 9.1. System specifications are listed in Table 9.1. A fraction of the laser load is absorbed by the optical elements via bulk volumetric absorption, resulting in thermal gradients in the lens assembly. The on-axis wavefront error, point spread function, modulation transfer function, and the change in focus is computed for laser powers of 40, 80, 120, 160, and 200 W. Nominally, each optical surface is spherical with surface curvatures listed in Table 9.2 and surface numbering shown in Fig. 9.2.

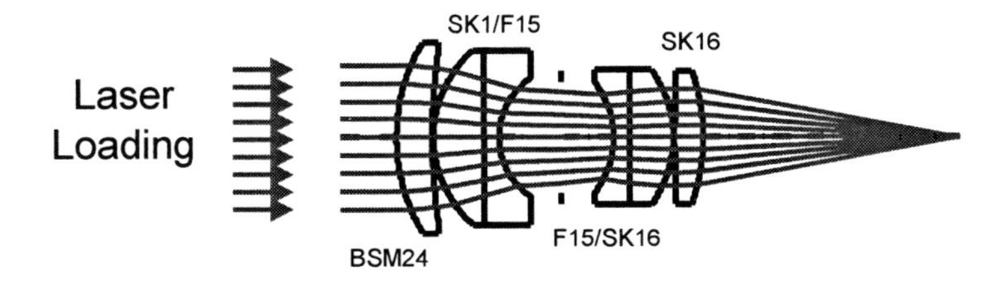

Figure 9.1 Double Gauss Lens Assembly*.

Table 9.1 Double Gauss lens assembly specifications.

OPTICAL SPECIFICATIONS	
Wavelength	587 nm
EPD	25 mm
f_{eff}	100 mm
$F/\#$	4.0
Housing	SS416

* U. S. Patent 2,532,751.

Table 9.2 Double Gauss optical surface curvatures.

NOMINAL SURFACE CURVATURE	
SURFACE	ρ
S1	0.017406
S2	0.005306
S3	0.028663
S4	Infinity
S5	0.046578
S6	−0.036989
S7	Infinity
S8	−0.028582
S9	0.001704
S10	−0.015844

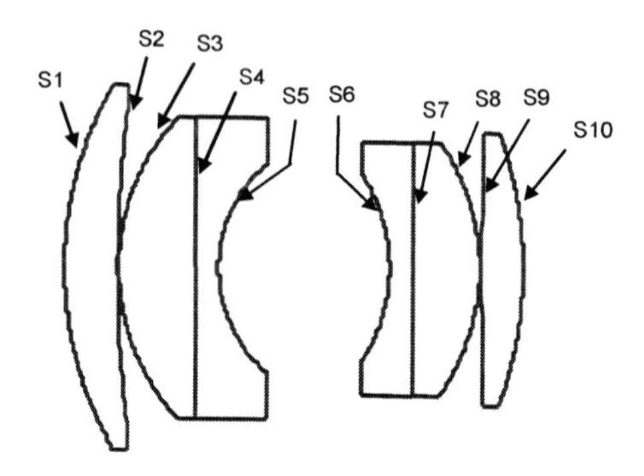

Figure 9.2 Double Gauss surface numbering.

9.2 Thermal Analysis

A thermal analysis was performed to compute the temperature distribution in the lens assembly due to the bulk volumetric absorption of the laser load. A two-step modeling effort was conducted using the thermal analysis software *Thermal Desktop. First, a heat-rate model was developed to compute the energy absorbed by the laser load. A heat source was defined to emit a parallel radiation flux and a mask was used to set the clear aperture for the lens assembly. Multiple surfaces of zero thickness with the appropriate indices of refraction were then defined for each lens element. The surface absorption coefficients were radially varied to account for the absorption characteristics of each glass and the energy distribution of the laser. The radiation flux was varied to yield the desired laser power. The resulting heat rates were then used in a steady-state thermal analysis to compute the temperature distribution for laser powers of 40, 80, 120, 160, and

* Thermal Desktop is a product of Cullimore and Ring Technologies, Inc., Littleton, CO.

200 W. The resulting temperature distribution is a radial gradient with a slight axial variation due to surface effects as shown in Fig. 9.3. The approximate radial gradient as a function of laser power is listed in Table 9.3. The temperatures were subsequently mapped to an MSC/Nastran finite element model using Thermal Desktop's shape function interpolation algorithm as demonstrated in Fig. 9.4.

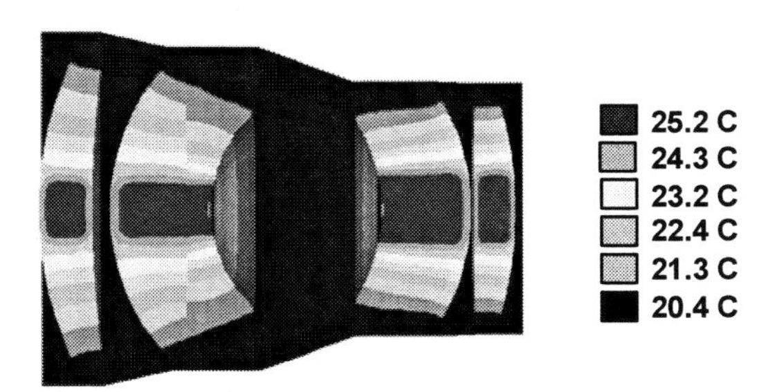

Figure 9.3 Temperature distribution due to laser loading of 200 watts.

Table 9.3 Radial gradient as a function of laser power.

LASER POWER (W)	RADIAL GRADIENT (°C)
40	1
80	2
120	3
160	4
200	5

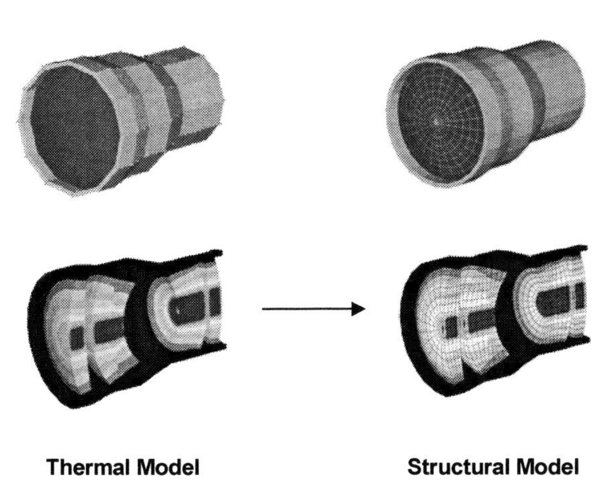

Figure 9.4 Temperature mapping using shape function interpolation.

9.3 Thermo-Elastic Analysis

A thermo-elastic analysis was performed using the finite element model to compute the rigid-body motions of the optical surfaces, the higher-order surface deformations, and the mechanical stress. The coefficient of thermal expansions for the optical glasses and housing materials are shown in Fig. 9.5. An exaggerated view of the resulting deformed shape is shown in Fig. 9.6. A least-squares fit to the surface deformations was performed to compute the rigid-body motions of each surface. Since the optical system and loading are rotationally symmetric this resulted in positional errors along the optical axis only. The higher-order surface deformations were fit to aspheric polynomials. The perturbed surface shapes are listed in Table 9.4 where ρ is the vertex curvature, and A, B, C, and D are the aspheric coefficients. The change in shape of the center surface for each of the two-cemented doublets is not given (surfaces 4 and 7). Typically, the index difference between the cemented glasses produces a negligible effect on system wavefront error. The higher-order surface departure is plotted as a function of radius in Fig. 9.7.

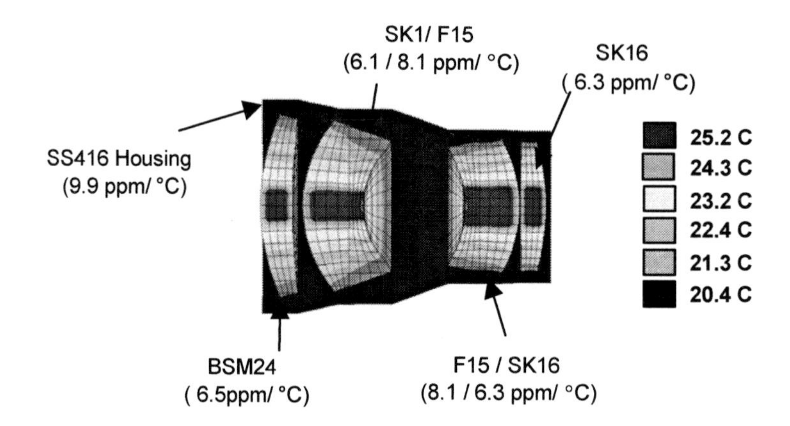

Figure 9.5 CTEs of optical glasses and housing.

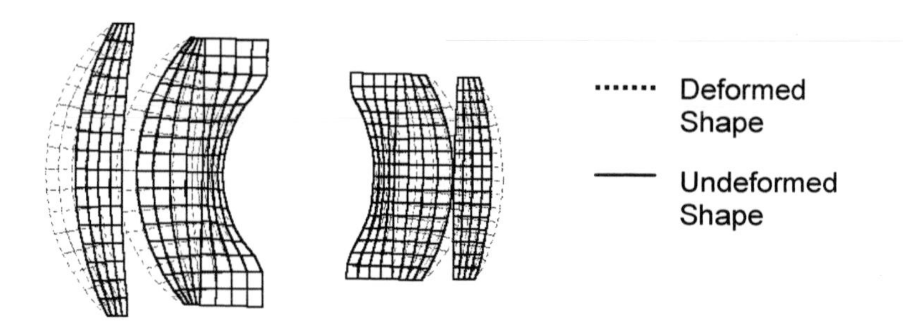

Figure 9.6 Double Gauss deformed and undeformed FEA results.

Table 9.4 Perturbed optical surface shape for 200 W laser load.

PERTURBED OPTICAL SURFACE SHAPE					
SURF	ρ	A	B	C	D
S1	0.017410	−4.2E-09	−3.8E-11	1.7E-14	−6.1E-17
S2	0.005302	1.5E-08	−1.8E-11	−2.2E-14	5.5E-17
S3	0.028673	−6.3E-09	−5.0E-11	−6.6E-16	−7.0E-17
S5	0.046570	3.6E-08	−1.1E-10	−4.1E-13	9.7E-16
S6	−0.036980	−3.2E-08	7.3E-11	1.0E-12	−6.0E-15
S8	−0.028593	9.7E-09	9.7E-11	−2.0E-13	1.2E-15
S9	0.001709	−1.5E-08	−8.2E-12	2.6E-13	−1.2E-15
S10	−0.015848	6.3E-09	6.5E-11	−2.8E-13	1.4E-15

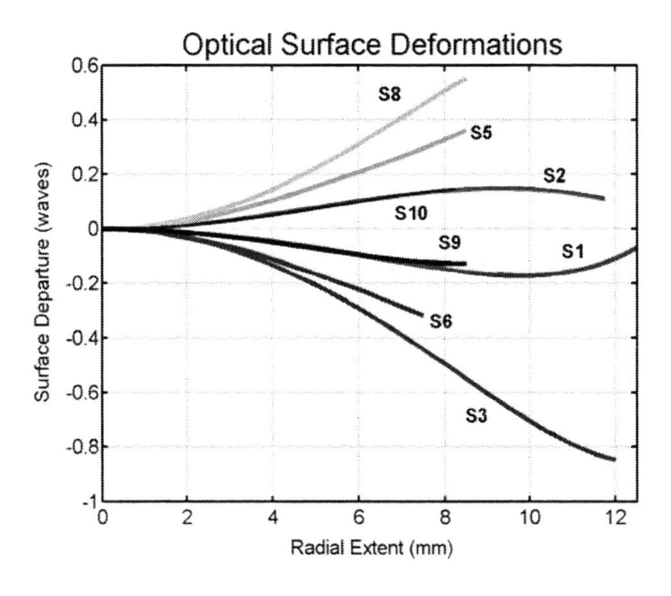

Figure 9.7 Optical surface departure.

9.4 Stress Birefringence Analysis

The laser load also produces stress in each of the optical elements, which creates wavefront error in the lens assembly. We assume for this specific example that polarization is not a concern. SigFit was used to trace a grid of rays through the stress field of each optical element. The OPD was incrementally summed for each ray producing an OPD map for each element. This data was subsequently fit to Zernike polynomials and formatted into CODE V wavefront interferogram files. The stress-optical coefficients for each of the glass types are listed in Table 9.5.

Table 9.5 Stress-optical coefficients.

STRESS-OPTICAL COEFFICIENTS ($\times 10^{-8}$ IN^2/LB$_F$)		
GLASS TYPE	$-K_{11}$	$-K_{12}$
BSM24	0.55	2.00
SK1	0.48	2.07
F15	1.65	3.65
SK16	0.69	1.92

9.5 Thermo-Optic Analysis

A thermo-optic analysis was performed to compute the wavefront error due to index changes. The relative thermo-optic coefficients are displayed with the temperature distribution for each of the optical elements in Fig. 9.8. SigFit computed OPD maps by tracing a grid of rays through each element and incrementally summing the OPD due to the temperature variation. The OPD maps were fit to Zernike polynomials and used to create wavefront interferogram files.

9.6 Optical Analysis

An optical analysis was performed using CODE V to compute optical performance as a function of laser power. For each laser load, the mechanical perturbations were applied to the optical model. Each optical surface was repostitioned along the optical axis, and the higher-order surface deformations were represented as aspheric surfaces. Mechanical stress and thermo-optic effects were accounted for in the optical model using wavefront interferogram files. On-axis performance of the lens assembly was then computed for laser loads of 40, 80, 120, 160, and 200 watts.

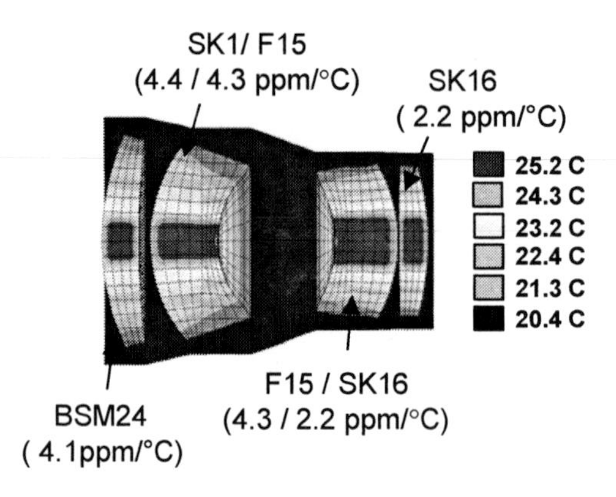

Figure 9.8 Thermo-optic coefficients for each glass.

The change in focus of the lens assembly, δ, as illustrated in Fig. 9.9, is listed as a function of laser power in Table 9.6. Peak-to-valley and rms wavefront error as a function of laser power is listed in tabular form using the dominant Zernike terms in Table 9.7 and shown graphically using interferogram plots in Fig. 9.10. In an interferogram plot, the wavefront error or OPD is calculated by the number of fringes. Each white and dark band represents half of a wave. As shown, the primary effect of the laser load is to put the system out of focus.

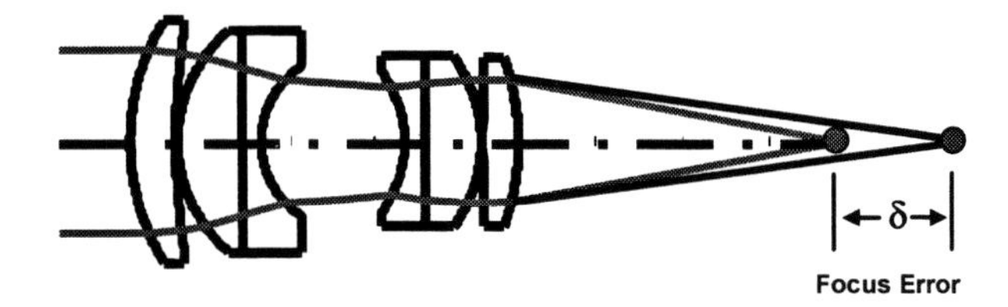

Figure 9.9 Laser loading induced focus error.

Table 9.6 Focus error.

FOCUS ERROR, δ	
LOAD CASE	ΔF (μM)
40 W	41
80 W	75
120 W	101
160 W	116
200 W	151

Table 9.7 Wavefront error fit to Zernike coefficients.

WAVEFRONT ERROR FRINGE ZERNIKE COEFFICIENTS					
LOAD CASE	PISTON	FOCUS	SPHERICAL	RMS	P-V
Nominal	0.56	0.82	0.25	0.48	1.6
40 W	0.92	1.09	0.16	0.63	2.2
80 W	1.10	1.32	0.21	0.76	2.6
120 W	1.38	1.50	0.11	0.84	3.9
160 W	1.46	1.59	0.12	0.92	3.2
200 W	1.91	1.83	−0.10	1.10	3.6

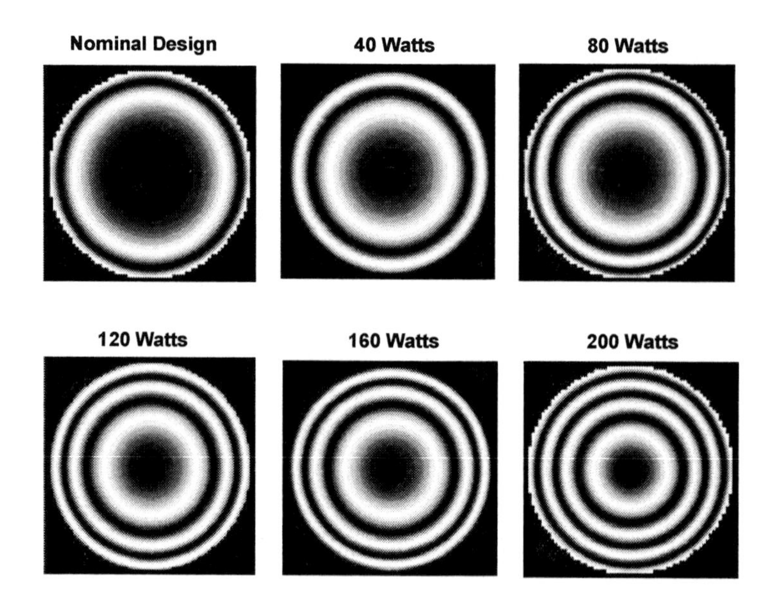

Figure 9.10 Interferogram plots of wavefront error as a function of laser loading.

A comparison of the wavefront error produced by the thermo-elastic deformations, the mechanical stress, and the thermo-optic effects for laser loading of 200 W is shown using interferogram plots in Fig. 9.11. The results indicate that, for this example, the thermo-elastic effects contribute approximately three times the wavefront error as the thermo-optic effects. The stress-induced wavefront error represents a small fraction of the total wavefront error.

The effect of the laser loading on the PSF and the MTF is shown in Figs. 9.12 and 9.13. Note as the laser power is increased, the blur diameter of the PSF increases and the MTF cutoff frequency decreases.

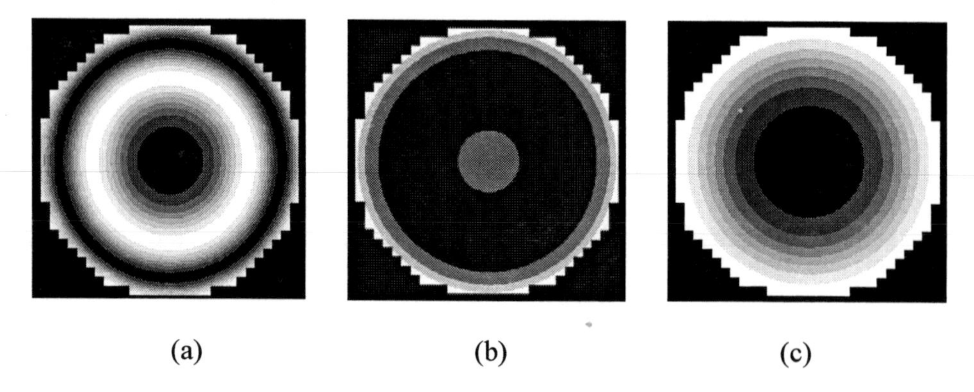

(a) (b) (c)

Figure 9.11 Contributions to system wavefront error using interferogram plots for 200 W load. (a) Thermo-elastic, (b) mechanical stress, and (c) thermo-optic effects.

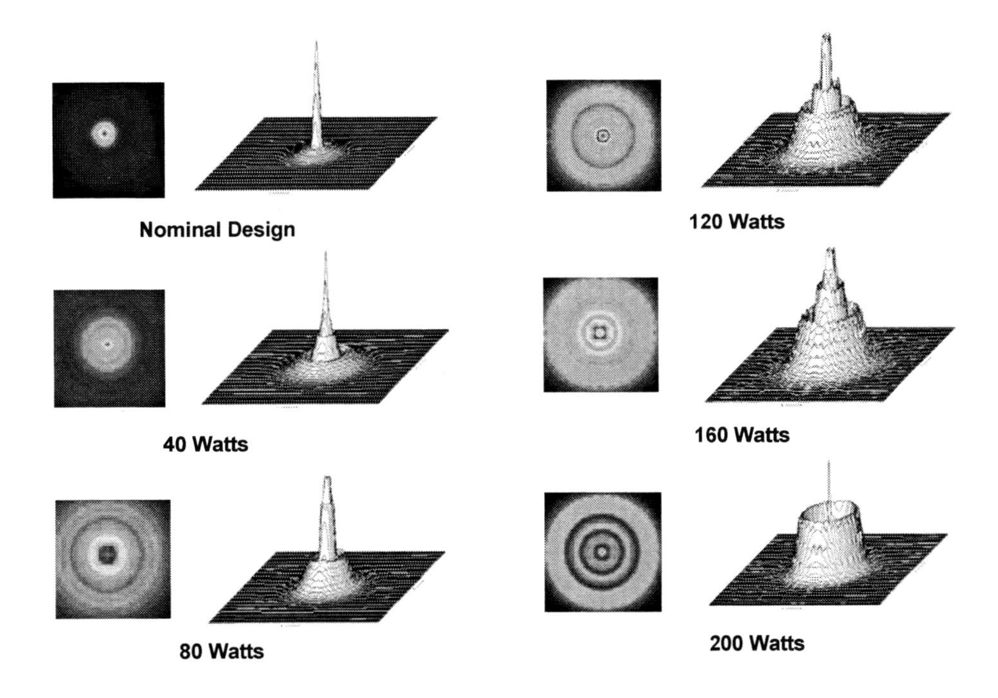

Figure 9.12 PSF as a function of laser loading.

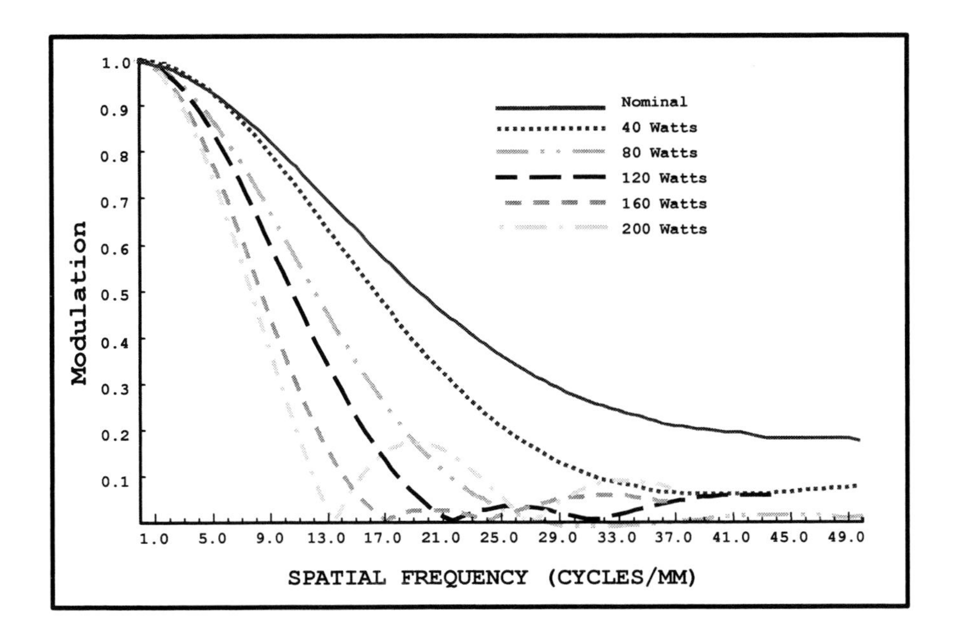

Figure 9.13 MTF as a function of laser loading.

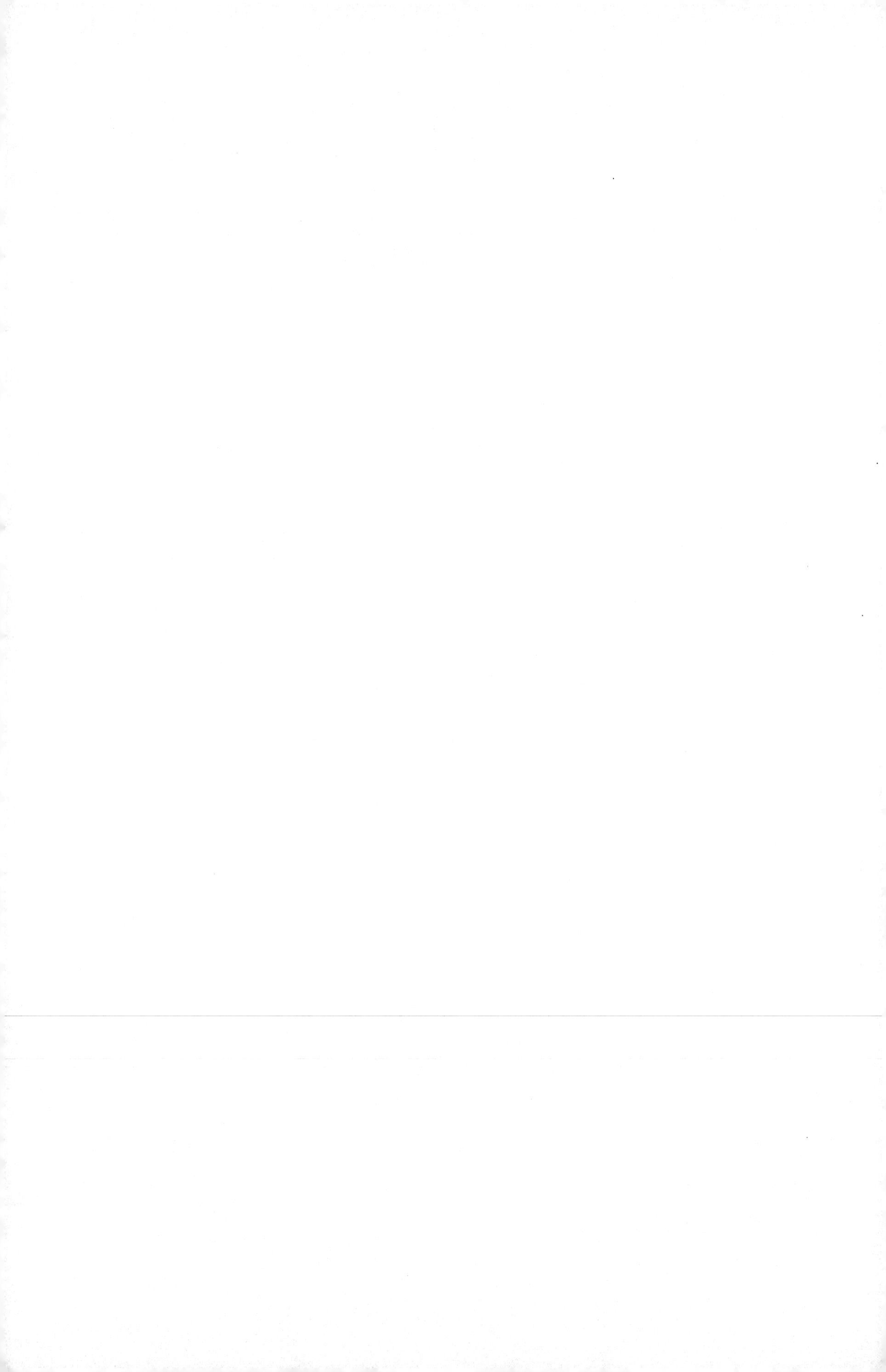

Index

Dr. Keith Doyle has over 15 years experience in the field of optical engineering, specializing in optomechanics and the multidisciplinary modeling of optical systems. He has worked on a diverse range of optical instruments including ground, aerial, and space-borne optics for astronomical and military applications, as well as optical systems for the microlithography, telecommunications, and consumer optics industries. He is currently employed as a Senior Systems Engineer at Optical Research Associates. Previous employers include Litton/Itek Optical Systems and MIT/Lincoln Laboratory. Dr. Doyle is an active organizer and participant in SPIE and OSA symposia. He provides technical reviews for Applied Optics, teaches short courses on a regular basis, and has authored and co-authored over 20 technical papers in optical engineering. He received his Ph.D. from the University of Arizona in Engineering Mechanics with a minor in the Optical Sciences in 1993.

Dr. Victor Genberg PE has over 35 years of experience in the application of finite element methods to high-performance optical structures, and is a recognized expert in optomechanics. He is currently President of Sigmadyne, Inc., an optical engineering consulting group that provides product design, development, and analysis to the optical community. Prior to starting Sigmadyne, Dr. Genberg worked at Eastman Kodak for 28 years serving as a technical specialist for commercial and military optical instruments. He is the author of SigFit, a commercially available software product for optomechanical analysis. Dr. Genberg is also a full professor (adjunct) of mechanical engineering at the University of Rochester where he teaches a variety of courses in finite elements, design, and optimization. He has over 30 publications, including two chapters in the CRC Handbook of Optomechanical Engineering. He received his Ph.D. from Case Western Reserve University in 1973.

Gregory Michels PE has worked for over 10 years in optomechanical design and analysis, and is currently Vice President of Sigmadyne, Inc. He specializes in computer-aided engineering, including finite element analysis, mathematical modeling, and design optimization as applied to high-performance optical instruments. Mr. Michels is also a technical contributor to Sigmadyne's product SigFit. Prior to joining Sigmadyne, he worked at Eastman Kodak for 5 years as a structural analyst on the Chandra X-Ray Observatory. Mr. Michels has authored or co-authored over 10 papers in this field and teaches short courses on finite element analysis and integrated modeling. He received his MS degree in mechanical engineering from the University of Rochester in 1994.